高等院校计算机任务驱动教改教材

移动互联网应用开发

（基于Android平台）

李维勇　主　编　杜亚杰　石　建　副主编

清华大学出版社
北京

内 容 简 介

本书是基于 Android Lollipop 平台进行移动互联网应用开发的入门级教程，通过众多开源案例项目，全面系统地介绍移动互联网应用开发的方法、技巧和模式。

全书共分为 9 章，从 Android 应用设计者的角度系统讲解了从事 Android 移动互联网应用开发必须要掌握的 Android 平台的相关技术和特性，主要内容包括数据流与数据解析，网络连接与管理、Android 中的 Socket 编程与 HTTP 编程，Web 应用编程、开放接口编程、Google 云服务技术等，全面总结了 Android 网络编程的基本原理、设计理念和设计模式，最后通过一个综合的案例项目阐述了 Android 移动互联网应用开发的方法和技巧。

本书以案例贯穿全程，知识结构清晰，语言简洁，易于读者学习和提高，非常适合初学 Android 移动互联网应用开发的在校大学生和希望系统掌握 Android 互联网编程的开发人员。

图书在版编目(CIP)数据

移动互联网应用开发：基于 Android 平台/李维勇主编. --北京：清华大学出版社，2016
高等院校计算机任务驱动教改教材
ISBN 978-7-302-42121-4

Ⅰ. ①移… Ⅱ. ①李… Ⅲ. ①移动终端－技术开发－高等学校－教材 Ⅳ. ①TN87

中国版本图书馆 CIP 数据核字(2015)第 267362 号

责任编辑：张龙卿
封面设计：徐日强
责任校对：刘 静
责任印制：宋 林

出版发行：清华大学出版社
网 址：http://www.tup.com.cn，http://www.wqbook.com
地 址：北京清华大学学研大厦 A 座 **邮 编**：100084
社 总 机：010-62770175 **邮 购**：010-62786544
投稿与读者服务：010-62776969，c-service@tup.tsinghua.edu.cn
质 量 反 馈：010-62772015，zhiliang@tup.tsinghua.edu.cn
课 件 下 载：http://www.tup.com.cn，010-62795764
印 刷 者：北京市人民文学印刷厂
装 订 者：三河市溧源装订厂
经 销：全国新华书店
开 本：185mm×260mm **印 张**：18.75 **字 数**：454 千字
版 次：2016 年 3 月第 1 版 **印 次**：2016 年 3 月第 1 次印刷
印 数：1～2500
定 价：36.00 元

产品编号：065950-01

编审委员会

出版说明

我国高职高专教育经过十几年的发展,已经转向深度教学改革阶段。教育部于2006年12月发布了教高[2006]第16号文件《关于全面提高高等职业教育教学质量的若干意见》,大力推行工学结合,突出实践能力培养,全面提高高职高专教学质量。

清华大学出版社作为国内大学出版社的领跑者,为了进一步推动高职高专计算机专业教材的建设工作,适应高职高专院校计算机类人才培养的发展趋势,根据教高[2006]第16号文件的精神,2007年秋季开始了切合新一轮教学改革的教材建设工作。该系列教材一经推出,就得到了很多高职院校的认可和选用,其中部分书籍的销售量超过了3万册。现重新组织优秀作者对部分图书进行改版,并增加了一些新的图书品种。

目前国内高职高专院校计算机网络与软件专业的教材品种繁多,但符合国家计算机网络与软件技术专业领域技能型紧缺人才培养培训方案,并符合企业的实际需要,能够自成体系的教材还不多。

我们组织国内对计算机网络和软件人才培养模式有研究并且有过一段实践经验的高职高专院校,进行了较长时间的研讨和调研,遴选出一批富有工程实践经验和教学经验的双师型教师,合力编写了这套适用于高职高专计算机网络、软件专业的教材。

本套教材的编写方法是以任务驱动、案例教学为核心,以项目开发为主线。我们研究分析了国内外先进职业教育的培训模式、教学方法和教材特色,消化吸收优秀的经验和成果。以培养技术应用型人才为目标,以企业对人才的需要为依据,把软件工程和项目管理的思想完全融入教材体系,将基本技能培养和主流技术相结合,课程设置重点突出、主辅分明、结构合理、衔接紧凑。教材侧重培养学生的实战操作能力,学、思、练相结合,旨在通过项目实践,增强学生的职业能力,使知识从书本中释放并转化为专业技能。

一、教材编写思想

本套教材以案例为中心,以技能培养为目标,围绕开发项目所用到的知识点进行讲解,对某些知识点附上相关的例题,以帮助读者理解,进而将知识转变为技能。

考虑到是以"项目设计"为核心组织教学，所以在每一学期配有相应的实训课程及项目开发手册，要求学生在教师的指导下，能整合本学期所学的知识内容，相互协作，综合应用该学期的知识进行项目开发。同时，在教材中采用了大量的案例，这些案例紧密地结合教材的各个知识点，循序渐进，由浅入深，在整体上体现了内容主导、实例解析、以点带面的模式，配合课程后期以项目设计贯穿教学内容的教学模式。

软件开发技术具有种类繁多、更新速度快的特点。本套教材在介绍软件开发主流技术的同时，帮助学生建立软件相关技术的横向及纵向的关系，培养学生综合应用所学知识的能力。

二、丛书特色

本系列教材体现目前工学结合的教改思想，充分结合教改现状，突出项目面向教学和任务驱动模式教学改革成果，打造立体化精品教材。

(1) 参照和吸纳国内外优秀计算机网络、软件专业教材的编写思想，采用本土化的实际项目或者任务，以保证其有更强的实用性，并与理论内容有很强的关联性。

(2) 准确把握高职高专软件专业人才的培养目标和特点。

(3) 充分调查研究国内软件企业，确定了基于 Java 和.NET 的两个主流技术路线，再将其组合成相应的课程链。

(4) 教材通过一个个教学任务或者教学项目，在做中学、学中做，以及边学边做，重点突出技能培养。在突出技能培养的同时，还介绍解决思路和方法，培养学生未来在就业岗位上的终身学习能力。

(5) 借鉴或采用项目驱动的教学方法和考核制度，突出计算机网络、软件人才培训的先进性、工具性、实践性和应用性。

(6) 以案例为中心，以能力培养为目标，并以实际工作的例子引入概念，符合学生的认知规律。语言简洁明了、清晰易懂，更具人性化。

(7) 符合国家计算机网络、软件人才的培养目标；采用引入知识点、讲述知识点、强化知识点、应用知识点、综合知识点的模式，由浅入深地展开对技术内容的讲述。

(8) 为了便于教师授课和学生学习，清华大学出版社正在建设本套教材的教学服务资源。清华大学出版社网站(www.tup.com.cn)免费提供教材的电子课件、案例库等资源。

高职高专教育正处于新一轮教学深度改革时期，从专业设置、课程体系建设到教材建设，依然是新课题。希望各高职高专院校在教学实践中积极提出意见和建议，并及时反馈给我们。清华大学出版社将对已出版的教材不断地修订、完善，提高教材质量，完善教材服务体系，为我国的高职高专教育继续出版优秀的高质量教材。

清华大学出版社

高职高专计算机任务驱动模式教材编审委员会

2014 年 3 月

前　言

国家“十二五”规划明确提出要加快发展新一代信息技术，移动互联网是我国未来发展战略的重点方向。移动终端 App 开发、移动互联网平台开发、基于 HTML5 的移动 Web 开发是移动互联网行业人才需求的三大方向。作为移动互联网应用开发工程师，基于 TCP/IP 的网络编程能力、终端轻量级数据库开发能力、基于典型公共云平台的应用开发能力是其必备的职业技能。

本书以 Android SDK Lollipop 5.0 为开发平台，以 Eclipse 为集成开发环境，结合笔者近年来在手机软件研发和教学中积累的经验，详细介绍 Android 平台移动互联网应用开发的相关知识。

本书共分为 9 章。第 1 章介绍 Java 中 I/O 数据流的基本知识，包括 Android 平台下字节流与字符流的处理，以及对象序列化与编码转换的技术。第 2 章介绍 XML 和 JSON 两种数据格式及其在 Android 中的解析方法，介绍了 JSON 解析工具 Gson 和 Json-lib 的使用方法。第 3 章介绍 Android 平台下网络连接与管理的方法，Wi-Fi 网络连接与管理的知识，以及网络服务优化的技术。第 4 章介绍网络编程的基本知识及相关 API，介绍了 Android 中 Socket 编程的方法和步骤，以及 UDP 编程和 NIO 编程的知识。第 5 章介绍 Android 平台中 HTTP 编程的方法，包括基于 HttpURLConnection 和 HttpClient 访问网络的方法，介绍了 Android-Async-Http 和 Volley 连接框架的使用技术。第 6 章介绍使用 WebView 构建 Web 应用的方法，以及使用 HTML5 开发 Web App 的方法。第 7 章介绍 Android 中 Web Service 编程的方法，并通过人人网等案例介绍基于开放平台实现互联网编程的技术。第 8 章介绍 Google 云开发技术，包括 Google 云备份技术、云信息技术和云存储技术等。第 9 章通过分析 Philm 项目的设计和实现过程，介绍移动互联网应用开发的技术架构与设计模式。

本书紧密结合初学者的学习习惯和认知规律，采用了大量简单而又实用的设计案例，使得读者在阅读时不会有障碍，并可通过简单的代码移植就能够生成新的应用。书中采用的开源案例项目把与 Android 开发相关的技术及设计完美结合，别具一格，弥补了一些 Android 设计人员知识的不足。

本书由李维勇担任主编，杜亚杰、石建担任副主编，李建林教授主审。本书的编写得到了南京信息职业技术学院、金审学院等院校的大力支持和帮助，杭州虹云网络科技有限公司为教材案例项目的策划、开发和测试提供了大量信息，清华大学出版社的编辑为本书的策划和出版提供了宝贵的经验和支持，在此表示衷心感谢。同时，本书在编写过程中，参考了大量的相关资料，吸取了许多同人的宝贵经验，在此一并致谢。

由于作者水平所限，疏漏难免，敬请广大读者提出宝贵意见和建议。教材配套课件、习题答案及源代码均可从清华大学出版社网站下载。

编　者

2016 年 1 月

目　录

第1章　数　据　流

移动互联网对网络数据流的处理要关注四个基本方面：数据流的编码、字节顺序、数据格式对应和取数。在编写网络程序时，必须注意这些问题，以使程序能正确处理通信的内容。

1.1　Java 中的 I/O

Java 中 I/O 操作主要是指使用 Java 进行输入、输出操作。Java 所有的 I/O 机制都是基于数据流进行输入输出的，这些数据流表示了字符或者字节数据的流动序列。Java 的 I/O 流提供了读写数据的标准方法。任何 Java 中表示数据源的对象都会提供以数据流的方式读写其数据的方法。

1.1.1　I/O 流

流是一组有顺序的、有起点和终点的字节集合，是对数据传输的总称或抽象。即数据在两设备间的传输称为流，流的本质是数据传输。当 Java 程序需要从数据源读取数据时，会开启一个到数据源的流。数据源可以是文件、内存或者网络等。同样，当程序需要输出数据到目的地时也一样会开启一个流，数据目的地也可以是文件、内存或者网络等。流的创建是为了更方便地处理数据的输入输出。

Java 根据数据传输特性将流抽象为各种类，以便于更直观地进行数据操作。流的常见分类方式包括两种。

(1) 根据处理数据类型的不同，分为字节流和字符流。

字节流也称为原始数据，需要用户读入后进行相应的编码转换。而字符流的实现是基于自动转换的，读取数据时会把数据按照 JVM 的默认编码自动转换成字符。

字节流由 java. io. InputStream 和 java. io. OutputStream 处理，而字符流由 java. io. Reader 和 java. io. Writer 处理。Reader 和 Writer 是 Java 后来版本新增加的处理类，以便使数据的处理更方便。

字节流和字符流的区别如下。

① 读写单位不同：字节流以字节(8bit)为单位，字符流以 16 位的 Unicode 码表示的字符为基本处理单位，根据码表映射字符，一次可能读多个字节。

② 处理对象不同：字节流能处理所有类型的数据(如图像、视频等)，而字符流只能处理字符类型的数据。

因此，在实际编程时，如果只是处理纯文本数据，就优先考虑使用字符流。除此之外都使用字节流。

(2) 根据数据流向不同,分为输入流和输出流。

程序从输入流读取数据源(图 1-1),程序向输出流写入数据(图 1-2)。对输入流只能进行读操作,对输出流只能进行写操作,程序中需要根据待传输数据的不同特性而使用不同的流。

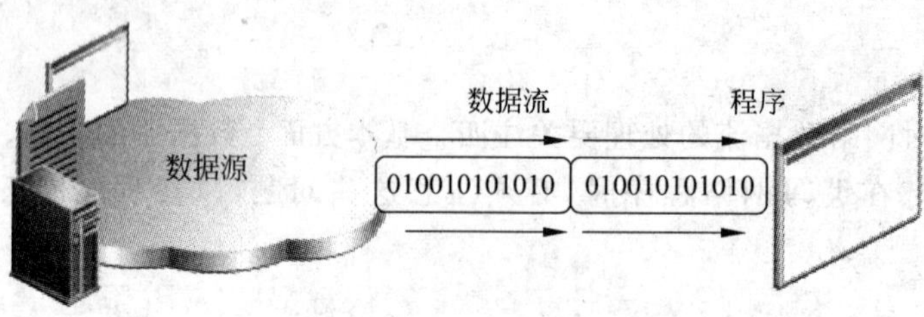

图 1-1 输入流

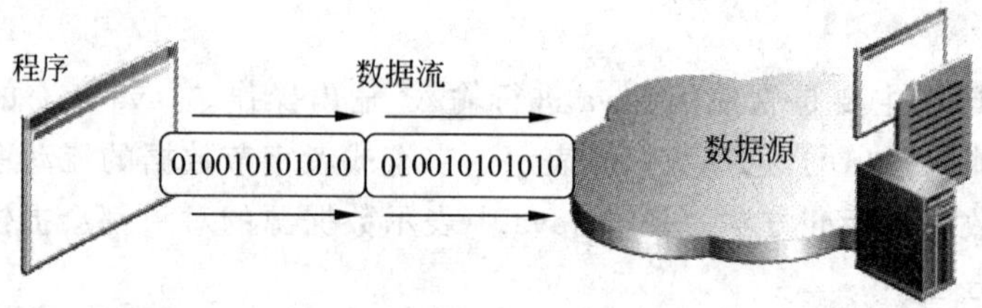

图 1-2 输出流

1.1.2 Java I/O 模型

Java 的 I/O 模型设计得非常优秀,它使用装饰者(Decorator)模式[①]来实现对各种输入输出流的封装。

Java I/O 主要包括如下三个部分。

(1) 流式部分:I/O 的主体部分。

(2) 非流式部分:主要包含一些辅助流式部分的类,如 File 类、RandomAccessFile 类和 FileDescriptor 等类。

(3) 其他类:文件读取部分与安全相关的类,如 SerializablePermission 类,以及与本地操作系统相关的文件系统的类,如 FileSystem 类、Win32FileSystem 类和 WinNTFileSystem 类。

对于流式部分,java.io 是大多数面向数据流的输入/输出类的主要软件包,按照功能可以分为 4 组。

- 基于字节操作的 I/O 接口:InputStream 和 OutputStream。
- 基于字符操作的 I/O 接口:Writer 和 Reader。
- 基于磁盘操作的 I/O 接口:File。
- 基于网络操作的 I/O 接口:Socket。

前两组主要是根据传输数据的数据格式,后两组主要是根据传输数据的方式。此外,

① 装饰者模式是面向对象编程领域中一种在不必改变原类文件和使用继承的情况下动态地往一个类中添加新行为的设计模式。

Java也对块传输提供支持，在核心库java.nio中采用的便是块I/O。流I/O的好处是简单易用，缺点是效率较低。块I/O效率很高，但编程比较复杂。

在Java中I/O操作的一般流程如下（以文件操作为例）：

① 使用File类打开一个文件；

② 通过字节流或字符流的子类，指定输出的位置；

③ 进行读/写操作；

④ 关闭输入/输出流。

提示： Java中的I/O操作属于资源操作，使用完毕后一定要记得关闭。

1.1.3　I/O异常

异常（Exception，又称为例外）是特殊的运行错误对象，对应着Java语言特定的运行错误处理机制。Java的异常类是处理运行时错误的特殊类，每一种异常类对应一种特定的运行错误。所有的Java异常类都是系统类库中Exception类的子类。

Java最常见的异常之一就是java.io.IOException，包括以下3种。

(1) public class EOFException

非正常到达文件尾或输入流尾时抛出的异常。

(2) public class FileNotFoundException

当文件找不到时抛出的异常。

(3) public class InterruptedIOException

当I/O操作被中断时抛出的异常。

1.2　字　节　流

字节流主要是操作byte类型数据，以byte数组为主，主要操作类包括InputStream和OutputStream。

1.2.1　InputStream

InputStream是所有的输入字节流的父类，它是一个抽象类，必须依靠其子类实现各种功能。继承自InputStream的流都是向程序中输入数据的，且数据单位为字节(8bit)。

1. 主要方法

InputStream提供的主要方法包括以下几种。

(1) public abstract int read() throws IOException

read()方法从输入流中读取一个byte的数据，返回0～255的int值。如果因为已经到达流末尾而没有可用的字节，则返回值为－1。在输入数据可用、检测到流末尾或者抛出异常前，此方法一直阻塞。

(2) public int read(byte[] b) throws IOException

read(byte[] b)从输入流中读取b.length个字节数据，并将其存储在缓冲区数组b中。返回值是读取的字节数。

(3) public int read(byte[] b, int off, int len) throws IOException

将输入流中最多 len 个数据字节读入偏移量为 off 的 b 数组中。但读取的字节也可能小于该值。以整数形式返回实际读取的字节数。

注意：

① 如果数组 b 的长度为 0,则不读取任何字节并返回 0；否则,尝试读取至少一个字节。

② 如果因为流位于文件末尾而没有可用的字节,则返回值为 −1；否则,至少读取一个字节并将其存储在 b 中。

③ read(byte[] b)方法将读取的第一个字节存储在元素 b[0]中,下一个存储在 b[1]中,其他依次类推。读取的字节数最多等于 b 的长度。设 k 为实际读取的字节数；这些字节将存储在 b[0]到 b[k−1]的元素中,不影响 b[k]到 b[b.length−1]的元素。

④ read(byte[] b, int off, int len)方法将读取的第一个字节存储在元素 b[off]中,下一个存储在 b[off+1]中,其他依次类推。读取的字节数最多等于 len。设 k 为实际读取的字节数；这些字节将存储在 b[off]到 b[off+k−1]的元素中,不影响 b[off+k]到 b[off+len−1]的元素。在任何情况下,b[0]到 b[off]的元素以及 b[off+len]到 b[b.length−1]的元素都不会受到影响。

⑤ read(byte[] b)和 read(byte[] b, int off, int len)这两个方法都是用来从流里读取多个字节的,但是这两个方法经常读取不到自己想要读取个数的字节。如 read(byte[] b)方法往往希望程序能读取到 b.length 个字节,而实际情况是,系统往往读取不了这么多。事实上,这个方法并不能保证读取这么多个字节,它只能最多读取这么多个字节(最少 1 个)。因此,如果要让程序读取 count 个字节(除非中途遇到 I/O 异常或者到了数据流的结尾),使用以下代码：

```
byte[ ] b = new byte[count];
int readCount = 0;
while (readCount < count) {
    readCount += in.read(bytes, readCount, count - readCount);
}
```

(4) public int available() throws IOException

返回输入流中可以读取的字节数,这个方法必须由继承 InputStream 类的子类对象调用才有用。在一次读取多个字节时,经常调用 available()方法,这个方法可以在读写操作前先得知数据流里有多少个字节可以读取。如果这个方法用在从本地文件读取数据时,一般不会遇到问题。但如果是用于网络操作,可能会遇到输入阻塞,当前线程将被挂起。如果 InputStream 对象调用这个方法,它只会返回 0。

为了保证调用 available()方法从网络获取 count 个数据,使用以下代码：

```
int count = 0;
while (count == 0) {
    count = in.available();
```

```
}
byte[] b = new byte[count];
in.read(b);
```

(5) public long skip(long n) throws IOException

跳过和丢弃此输入流中数据的 n 个字节。出于各种原因，skip()方法结束时跳过的字节数可能小于该数，也可能为 0。导致这种情况的原因很多，跳过 n 个字节之前已到达文件末尾只是其中一种可能。返回跳过的实际字节数。如果 n 为负值，则不跳过任何字节。

(6) public void close() throws IOException

关闭流并释放内存资源。

在 Android 中，存放在 Assets 目录(存放路径是 project/assets/file.name)中的文件会被原封不动地拷贝到 APK 中，而不会像其他资源文件那样被编译成二进制的形式。下面的代码演示了使用 InputStream 读取 Assets 中文件的方法。

```
public class InputStreamActivity extends Activity {
    private String fileName = "InputStreamExample.txt";

    private TextView mTextView;

    @Override
    protected void onCreate(Bundle savedInstanceState) {
        super.onCreate(savedInstanceState);
        setContentView(R.layout.activity_input_stream);

        mTextView = (TextView) findViewById(R.id.tv);
        mTextView.setText(getFromAsset(fileName));
    }

    private String getFromAsset(String fileName) {
        String res = "";
        try {
            InputStream in = getResources().getAssets().open(fileName);
            int length = in.available();
            byte[] buffer = new byte[length];
            in.read(buffer);
            res = EncodingUtils.getString(buffer, "GB2312");
        } catch (Exception e) {
            e.printStackTrace();
        }
        return res;
    }

}
```

2. 主要子类

ByteArrayInputStream、StringBufferInputStream、FileInputStream 是三种基本的介质

流,它们分别从 Byte 数组、StringBuffer 和本地文件中读取数据。

其他的输入流处理类都是装饰类(Decorator 模式),包括以下几类。

- BufferedInputStream:提供缓冲功能。
- DataInputStream:允许应用程序以与机器无关方式从底层输入流中读取基本 Java 数据类型。应用程序可以使用数据输出流写入稍后由数据输入流读取的数据。
- PipedInputStream:允许从与其他线程共用的管道中读取数据。当连接到一个 PipedOutputStream 后,它会读取后者输出到管道的数据。
- PushbackInputStream:允许放回已经读取的数据。
- SequenceInputStream:能对多个 InputStream 进行顺序处理。
- ObjectInputStream 和所有 FilterInputStream 的子类:都是装饰流(装饰器模式的主角)。

InputStream 的继承关系如图 1-3 所示。

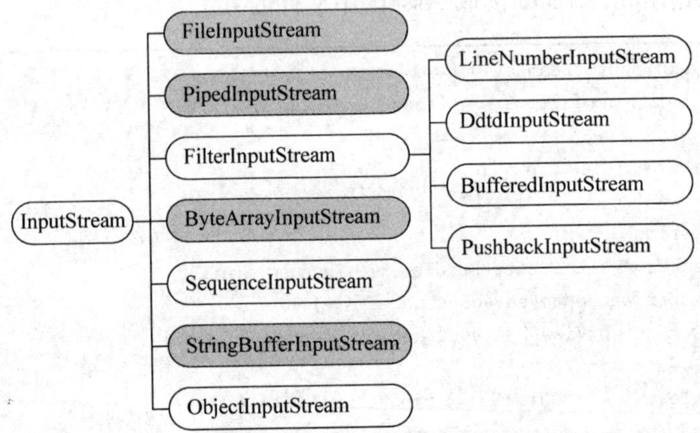

图 1-3 InputStream 的继承关系

下面的代码演示了使用 FileInputStream 和 FileOutputStream 读写 SD 卡文件的方法。

```
public class FileIOActivity extends Activity {

    private String filePath;
    private String fileName = "fileio.txt";

    private TextView mTextView;

    @Override
    protected void onCreate(Bundle savedInstanceState) {
        super.onCreate(savedInstanceState);
        setContentView(R.layout.activity_file_io);

        mTextView = (TextView) findViewById(R.id.tv);

        //监听 SD 卡是否被加载
        if (!Environment.getExternalStorageState().equals(
```

```
                    android.os.Environment.MEDIA_MOUNTED)) {
                Toast.makeText(getApplicationContext(), "请插入 SD 卡",
                        Toast.LENGTH_SHORT).show();
                finish(); //如果没有加载 sd 卡,则退出应用
            } else {
                //获取 SD 卡根路径
                filePath = Environment.getExternalStorageDirectory()
                        .getAbsolutePath();
                String info = "白日依山尽,\n 黄河入海流。\n 欲穷千里目,\n 更上一层楼。";

                try {
                    writeSDFile(fileName, info);
                } catch (IOException e) {
                    e.printStackTrace();
                    mTextView.setText("文件写入失败");
                    return;
                }
            }

            try {
                mTextView.setText(readSDFile(fileName));
            } catch (IOException e) {
                e.printStackTrace();
                mTextView.setText("文件读取失败");
            }

        }

    //读文件
    public String readSDFile(String fileName) throws IOException {
        String res = "";

        File file = new File(filePath, fileName);
        FileInputStream fis = new FileInputStream(file);
        int length = fis.available();
        byte[] buffer = new byte[length];
        fis.read(buffer);
        res = EncodingUtils.getString(buffer, "UTF-8");
        fis.close();

        return res;
    }

    //写文件
    public void writeSDFile(String fileName, String write_str)
            throws IOException {
        File file = new File(filePath, fileName);
        FileOutputStream fos = new FileOutputStream(file);
        byte[] bytes = write_str.getBytes();
```

```
            fos.write(bytes);
            fos.close();
        }

}
```

注意：需要在 AndroidManifest. xml 文件中加入如下权限：

```
<uses-permission android:name="android.permission.WRITE_EXTERNAL_STORAGE" />
```

1.2.2 OutputStream

OutputStream 是所有输出字节流的父类，它是一个抽象类。

1. 主要方法

OutputStream 提供的重要方法包括以下几种。

(1) public abstract void write(int b) throws IOException

先将 int 类型的 b 转换为 byte 类型，然后写入 b 的 8 个低位，b 的 24 个高位将被忽略。

(2) public void write(byte[] b) throws IOException

将 b. length 个字节从指定的 byte 数组写入此输出流。write(b)方法应该与调用 write (b, 0, b. length)的效果完全相同。

(3) public void write(byte[] b, int off, int len) throws IOException

将指定 byte 数组中从偏移量 off 开始的 len 个字节写入此输出流。write(b, off, len)的常规协定是：将数组 b 中的某些字节按顺序写入输出流；元素 b[off]是此操作写入的第一个字节，b[off+len-1]是此操作写入的最后一个字节。

说明：

① 如果参数 b 为 null，则抛出 NullPointerException。

② 如果参数 off 为负，或 len 为负，或者 off+len 大于数组 b 的长度，则抛出 IndexOutOfBoundsException。

(4) public void flush() throws IOException

刷新此输出流并强制写出所有缓冲的输出字节。flush 的常规协定是：如果此输出流的实现已经缓冲了以前写入的任何字节，则调用此方法指示应将这些字节立即写入它们预期的目标。

说明：如果此流的预期目标是由基础操作系统提供的一个抽象(如一个文件)，则刷新此流只能保证将以前写入流的字节传递给操作系统进行写入，但不保证能将这些字节实际写入物理设备(如磁盘驱动器)。

OutputStream 的 flush()方法不执行任何操作。

(5) publicvoid close() throws IOException

关闭流并释放内存资源。关闭的流不能执行输出操作，也不能重新打开。

2. 主要子类

ByteArrayOutputStream、FileOutputStream 是两种基本的介质流，前者实现了一个输出流，其中的数据被写入一个 byte 数组；后者实现了把数据流写入文件的功能。缓冲区会随着数据的不断写入而自动增长。可使用 toByteArray()和 toString()获取数据。

其他的输出流处理类都是装饰类(Decorator 模式),包括以下几种。

- BufferedOutputStream：提供缓冲功能的输出流,在写出完成之前要调用 flush()来保证数据的输出。
- DataOutputStream：数据输出流允许应用程序以适当方式将基本 Java 数据类型写入输出流中。然后,应用程序可以使用数据输入流将数据读入。
- PipedOutputStream：允许以管道的方式来处理流。可以将管道输出流连接到管道输入流来创建通信管道。管道输出流是管道的发送端。通常,数据由某个线程写入 PipedOutputStream 对象,并由其他线程从连接的 PipedInputStream 读取。
- PrintStream：为其他输出流添加了功能,使它们能够方便地打印各种数据值表示形式。Java 经常用到的 System. out 或者 System. err 都是 PrintStream。

OutputStream 的继承关系如图 1-4 所示。

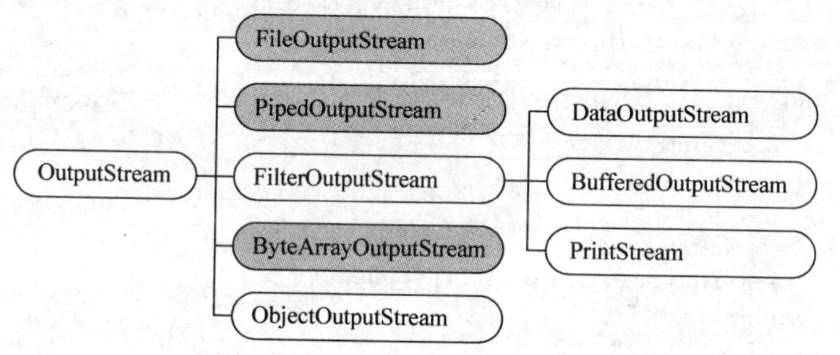

图 1-4　OutputStream 的继承关系

Android 可以直接将文件保存到设备的内部存储器上。默认情况下,保存到内部存储器上的文件对应用程序是私有的,其他应用无法访问它们。当用户卸载应用程序时,这些文件会被移除。

私有文件的存放路径是 data/data/[PACKAGE_NAME]/files/file. name,如图 1-5 所示。

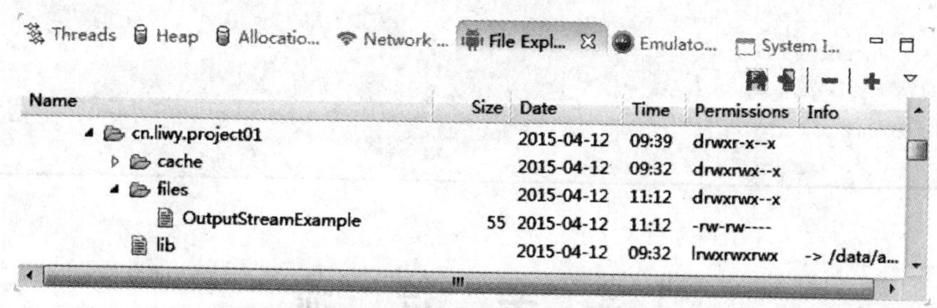

图 1-5　私有文件的存放路径

下面的示例演示了使用 FileInputStream 和 FileOutputStream 读写私有文件的方法。

```
public class OutputStreamActivity extends Activity {
    private String fileName = "OutputStreamExample";

```

```
    private TextView mTextView;

    @Override
    protected void onCreate(Bundle savedInstanceState) {
        super.onCreate(savedInstanceState);
        setContentView(R.layout.activity_output_stream);

        mTextView = (TextView) findViewById(R.id.tv);

        writeFileData(fileName, "欢迎学习基于 Android 平台的移动互联网编程");
        mTextView.setText(readFileData(fileName));
    }

    private void writeFileData(String fileName, String message) {
        try {
            FileOutputStream fout = openFileOutput(fileName, MODE_PRIVATE);
            byte[] bytes = message.getBytes();
            fout.write(bytes);
            fout.close();
        } catch (Exception e) {
            e.printStackTrace();
        }
    }

    private String readFileData(String fileName) {
        String res = "";
        try {
            FileInputStream fin = openFileInput(fileName);
            int length = fin.available();
            byte[] buffer = new byte[length];
            fin.read(buffer);
            res = EncodingUtils.getString(buffer, "UTF-8");
            fin.close();
        } catch (Exception e) {
            e.printStackTrace();
        }

        return res;
    }

}
```

1.3 字符流

字符流主要操作类包括 Reader 和 Writer。

1.3.1 Reader

Reader 是所有的输入字符流的父类,它是一个抽象类。

1. 主要方法

Reader 提供的重要方法包括以下几种。

(1) public int read() throws IOException

读取一个字符并以整数的形式返回(0~255)。如果返回-1,则表示已经到输入流的末尾。在字符可用、发生 I/O 错误或者已到达流的末尾前,此方法一直阻塞。

(2) public int read(char [] cbuf) throws IOException

读取一系列字符并存储到一个数组 buffer,返回实际读取的字符数。如果返回-1,则表示已经到输入流的末尾。

(3) public int read(char [] cbuf, int offset, int length) throws IOException

从 offset 位置开始,读取 length 个字符并存储到一个数组 buffer,返回实际读取的字符数,如果返回-1,则表示已经到输入流的末尾。

(4) public long skip(long n) throws IOException

跳过 *n* 个字符不读,返回实际跳过的字符数。

(5) public void close() throws IOException

关闭流并释放内存资源。

2. 主要子类

与数据源直接接触的类包括以下几种。

- CharArrayReader: 从内存中的字符数组读入数据,以对数据进行流式读取。
- StringReader: 从内存中的字符串读入数据,以对数据进行流式读取。
- FileReader: 从文件中读入数据。注意这里读入数据时会根据 JVM 的默认编码对数据进行内转换,而不能指定使用的编码。所以当文件使用的编码不是 JVM 默认编码时,不要使用这种方式。要正确地转码,使用 InputStreamReader。

装饰类包括以下几种。

- BufferedReader: 提供缓冲功能,使用 readLine()方法读取行。
- LineNumberReader: 提供读取行的控制,如 getLineNumber()等方法。
- InputStreamReader: 字节流通向字符流的桥梁,它使用指定的 charset 读取字节并将其解码为字符。

Reader 的继承关系如图 1-6。

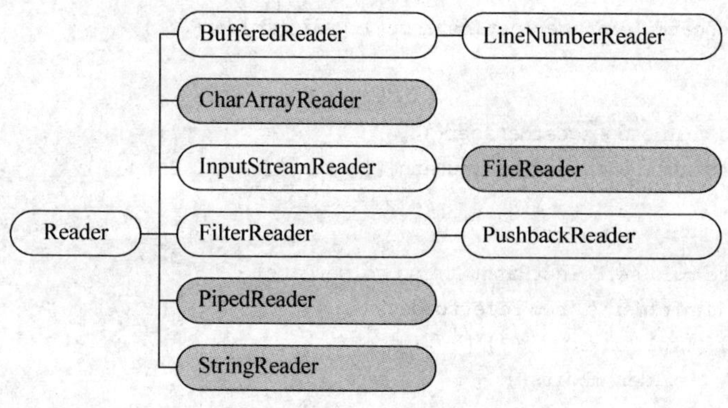

图 1-6 Reader 的继承关系

在调试程序的时候,一般都是在 logcat 中查看日志信息,以便找出 BUG 和调试信息,但是如果在真正的计算机上运行,就无法查看 LogCat 窗口。下面的工具类 LogcatFileManager 使用 BufferedReader 和 InputStreamReader 读取日志信息并保存到 SD 卡上,以便调试程序时跟踪查看。

```
public class LogcatFileManager {
    private static LogcatFileManager INSTANCE = null;
    private static String PATH_LOGCAT;
    private LogDumper mLogDumper = null;
    private int mPId;
    private SimpleDateFormat simpleDateFormat1 = new SimpleDateFormat(
            "yyyyMMdd");
    private SimpleDateFormat simpleDateFormat2 = new SimpleDateFormat(
            "yyyy-MM-dd HH:mm:ss");

    public static LogcatFileManager getInstance() {
        if (INSTANCE == null) {
            INSTANCE = new LogcatFileManager();
        }
        return INSTANCE;
    }

    private LogcatFileManager() {
        mPId = android.os.Process.myPid();
    }

    public void startLogcatManager(Context context) {
        String folderPath = null;
        if (Environment.getExternalStorageState().equals(
                Environment.MEDIA_MOUNTED)) {
            folderPath = Environment.getExternalStorageDirectory()
                    .getAbsolutePath() + File.separator + "MMF-Logcat";
        } else {
            folderPath = context.getFilesDir().getAbsolutePath()
                    + File.separator + "MMF-Logcat";
        }
        LogcatFileManager.getInstance().start(folderPath);
    }

    public void stopLogcatManager() {
        LogcatFileManager.getInstance().stop();
    }

    private void setFolderPath(String folderPath) {
        File folder = new File(folderPath);
        if (!folder.exists()) {
            folder.mkdirs();
        }
        if (!folder.isDirectory()) {
```

```
            throw new IllegalArgumentException(
                    "The logcat folder path is not a directory: " + folderPath);
        }

        PATH_LOGCAT = folderPath.endsWith("/") ? folderPath : folderPath + "/";
        LogUtils.d(PATH_LOGCAT);
    }

    public void start(String saveDirectoy) {
        setFolderPath(saveDirectoy);
        if (mLogDumper == null) {
            mLogDumper = new LogDumper(String.valueOf(mPId), PATH_LOGCAT);
        }
        mLogDumper.start();
    }

    public void stop() {
        if (mLogDumper != null) {
            mLogDumper.stopLogs();
            mLogDumper = null;
        }
    }

    private class LogDumper extends Thread {
        private Process logcatProc;
        private BufferedReader mReader = null;
        private boolean mRunning = true;
        String cmds = null;
        private String mPID;
        private FileOutputStream out = null;

        public LogDumper(String pid, String dir) {
            mPID = pid;
            try {
                out = new FileOutputStream(new File(dir, "logcat - "
                        + simpleDateFormat1.format(new Date()) + ".log"), true);
            } catch (FileNotFoundException e) {
                e.printStackTrace();
            }

            /**
             * log level: *:v , *:d , *:w , *:e , *:f , *:s
             * Show the current mPID process level of E and W log.
             */
            cmds = "logcat *:e *:w | grep \"(" + mPID + ")\"";
        }

        public void stopLogs() {
            mRunning = false;
```

```
        }

        @Override
        public void run() {
            try {
                logcatProc = Runtime.getRuntime().exec(cmds);
                mReader = new BufferedReader(new InputStreamReader(
                        logcatProc.getInputStream()), 1024);
                String line = null;
                while (mRunning && (line = mReader.readLine()) != null) {
                    if (!mRunning) {
                        break;
                    }
                    if (line.length() == 0) {
                        continue;
                    }
                    if (out != null && line.contains(mPID)) {
                        out.write((simpleDateFormat2.format(new Date()) + " "
                                + line + "\n").getBytes());
                    }
                }
            } catch (IOException e) {
                e.printStackTrace();
            } finally {
                if (logcatProc != null) {
                    logcatProc.destroy();
                    logcatProc = null;
                }
                if (mReader != null) {
                    try {
                        mReader.close();
                        mReader = null;
                    } catch (IOException e) {
                        e.printStackTrace();
                    }
                }
                if (out != null) {
                    try {
                        out.close();
                    } catch (IOException e) {
                        e.printStackTrace();
                    }
                    out = null;
                }
            }
        }

    }
}
```

在 Activity 中启动和停止 LogcatFileManager 的方法如下：

```
public class ReaderActivity extends Activity {

    private String filePath;

    private TextView mTextView;

    @Override
    protected void onCreate(Bundle savedInstanceState) {
        super.onCreate(savedInstanceState);
        setContentView(R.layout.activity_reader);

        mTextView = (TextView) findViewById(R.id.tv);

        //监听 SD 卡是否被加载
        if (!Environment.getExternalStorageState().equals(
                android.os.Environment.MEDIA_MOUNTED)) {
            Toast.makeText(getApplicationContext(), "请插入 SD 卡",
                    Toast.LENGTH_SHORT).show();
            finish(); //如果没有加载 SD 卡,则退出应用
        } else {
            //获取 SD 卡根路径
            filePath = Environment.getExternalStorageDirectory()
                    .getAbsolutePath();
            LogcatFileManager.getInstance().start(filePath);
            mTextView.setText("正在将日志信息写入 SD 卡");
        }

    }

    @Override
    protected void onStop() {
        super.onStop();
        LogcatFileManager.getInstance().stop();
    }

}
```

注意：需要在 AndroidManifest.xml 文件中加入如下权限：

```
<uses-permission android:name="android.permission.WRITE_EXTERNAL_STORAGE" />
<uses-permission android:name="android.permission.READ_LOGS" />
```

1.3.2 Writer

Writer 是所有的输出字符流的父类，它是一个抽象类。

1. 主要方法

Writer 提供的重要方法包括以下几种。

(1) public void write(int c) throws IOException

向输出流中写入一个字符数据,要写入的字符包含在给定整数值的 16 个低位中,16 个高位被忽略。

(2) public void write(char [] cbuf) throws IOException

将一个字符类型数组中的数据写入输出流。

(3) public void write(char [] cbuf, int offset, int length) throws IOException

将一个字符类型数组中的从指定位置(offset)开始的 length 个字符写入输出流。

(4) public void write(String string) throws IOException

将一个字符串中的字符输入输出流。

(5) public void write(String string, int offset, int length) throws IOException

将一个字符串从指定位置(offset)开始的 length 个字符写入输出流。

(6) void flush() throws IOException

将输出流中缓冲的数据全部输出到目的地。

(7) public void close() throws IOException

关闭流并释放内存资源。

2. 主要子类

与数据目的相关的类包括以下几种。

- CharArrayWriter:把内存中的字符数组写入输出流,输出流的缓冲区会自动增加大小。输出流的数据可以通过一些方法重新获取。
- StringWriter:一个字符流,可以用缓冲区中的字符串输出。
- FileWriter:把数据写入文件。

装饰类包括以下几种。

- BufferedWriter:提供缓冲功能。
- OutputStreamWriter:字符流通向字节流的桥梁。可使用指定的 charset 将要写入流中的字符编码成字节。
- PrintWriter:向文本输出流打印对象的格式化表示形式。

Writer 的继承关系如图 1-7 所示。

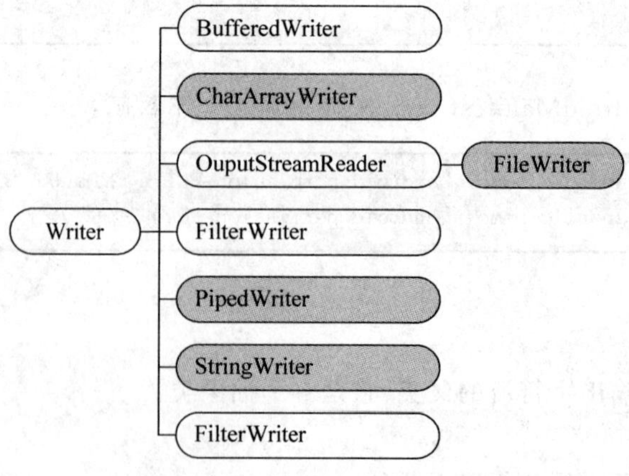

图 1-7 Writer 的继承关系

下面的服务 LogService 改进了 LogcatFileManager 的不足，其中写日志信息的部分代码如下：

```
public class LogService extends Service {
    private OutputStreamWriter writer;
    ...
    private void recordLogServiceLog(String msg) {
        if (writer != null) {
            try {
                Date time = new Date();
                writer.write(myLogSdf.format(time) + " : " + msg);
                writer.write("\n");
                writer.flush();
            } catch (IOException e) {
                e.printStackTrace();
                Log.e(TAG, e.getMessage(), e);
            }
        }
    }
}
```

1.4　对象序列化与编码转换

当两个进程在进行远程通信时，彼此可以发送各种类型的数据。无论是何种类型的数据，都会以二进制序列的形式在网络上传送。发送方需要把这个对象序列化为字节序列，才能在网络上传送；接收方则需要通过反序列化把字节序列再恢复为被处理的对象。在通信的同时还需要关注对象的编码方式。

1.4.1　对象序列化

1. 相关 API

java.io.ObjectOutputStream 代表对象输出流，它的 writeObject(Object obj)方法可对参数指定的 obj 对象进行序列化，把得到的字节序列写到一个目标输出流中。

java.io.ObjectInputStream 代表对象输入流，它的 readObject()方法从一个源输入流中读取字节序列，再把它们反序列化为一个对象，并将其返回。

java.io.Serializable 接口对于要实现它的类来说，主要用来通知 Java 虚拟机，需要将一个对象序列化。

注意：只有实现了 Serializable 和 Externalizable 接口的类的对象才能被序列化。Externalizable 接口继承自 Serializable 接口，实现 Externalizable 接口的类完全由自身来控制序列化的行为，而仅实现 Serializable 接口的类可以采用默认的序列化方式。

2. 序列化的一般步骤

对象序列化包括如下步骤：

(1) 创建一个对象输出流,它可以包装一个其他类型的目标输出流,如文件输出流;

(2) 通过对象输出流的 writeObject()方法写入对象。

对象反序列化的步骤如下:

(1) 创建一个对象输入流,它可以包装一个其他类型的源输入流,如文件输入流;

(2) 通过对象输入流的 readObject()方法读取对象。

注意:

① 序列化时,只对对象的状态进行保存,而不管对象的方法。

② 当一个父类实现序列化,子类自动实现序列化,不需要显示实现 Serializable 接口。

③ 当一个对象的实例变量引用其他对象,序列化该对象时,也把引用对象进行序列化。

④ 出于安全方面的原因以及资源分配等方面的考虑,并非所有的对象都可以序列化。

下面的示例演示了保存序列化数据到 SD 卡并显示的方法。

```
public class SerializableActivity extends Activity {

    private String fileName = "SerializableExample.txt";
    private String filePath;

    private TextView mTextView;

    @Override
    protected void onCreate(Bundle savedInstanceState) {
        super.onCreate(savedInstanceState);
        setContentView(R.layout.activity_serializable);

        filePath = Environment.getExternalStorageDirectory()
                .getAbsolutePath();

        mTextView = (TextView) findViewById(R.id.tv);

        saveSerializableData();

        try {
            mTextView.setText(readSerializableData());
        } catch (StreamCorruptedException e) {
            e.printStackTrace();
        } catch (ClassNotFoundException e) {
            e.printStackTrace();
        } catch (IOException e) {
            e.printStackTrace();
        }
    }

    private void saveSerializableData() {

        ChainList<Company> list = new ChainList<Company>();
```

```
        Company company1 = new Company();
        company1.setId("00001");
        company1.setName("10086");
        company1.setPwd("88881234");

        Company company2 = new Company();
        company2.setId("00002");
        company2.setName("10000");
        company2.setPwd("12345678");

        list.addObject(company1).addObject(company2);

        Serialize2SDcard.serialize2SDcard(list, filePath + File.separator
                + fileName);
    }

    private String readSerializableData() throws StreamCorruptedException,
            IOException, ClassNotFoundException {

        FileInputStream fis = new FileInputStream(filePath + File.separator
                + fileName);
        ObjectInputStream ois = new ObjectInputStream(fis);
        @SuppressWarnings("unchecked")
        ChainList<Company> list = (ChainList<Company>) ois.readObject();

        StringBuilder sb = new StringBuilder();

        for (Company company : list) {

            sb.append(company.toString() + "\n");
        }

        return sb.toString();
    }
}

class Serialize2SDcard {

    public static void serialize2SDcard(Object target, String file) {

        FileOutputStream fos = null;
        ObjectOutputStream o = null;
        try {
            fos = new FileOutputStream(file);
            o = new ObjectOutputStream(fos);
            o.writeObject(target);

        } catch (FileNotFoundException e) {
            e.printStackTrace();
```

```
            System.out.println("没有找到文件!");
        } catch (IOException e) {
            e.printStackTrace();
        } finally {
            try {
                fos.close();
                o.close();
            } catch (IOException e) {
                e.printStackTrace();
            }

        }

    }

    @SuppressWarnings("resource")
    public static Object getObjectFromSDcard(String file) {

        if (!new File(file).exists())
            return null;

        FileInputStream fis = null;
        ObjectInputStream in = null;
        try {
            fis = new FileInputStream(file);
            in = new ObjectInputStream(fis);

            return in.readObject();

        } catch (FileNotFoundException e) {
            e.printStackTrace();
            System.out.println("没有找到文件!");
        } catch (StreamCorruptedException e) {
            e.printStackTrace();
        } catch (IOException e) {
            e.printStackTrace();
        } catch (ClassNotFoundException e) {
            e.printStackTrace();
        } finally {

            try {
                fis.close();
                in.close();
            } catch (IOException e) {
                e.printStackTrace();
            }
        }

        return null;
```

```
    }

}
```

说明：虚拟机是否允许对象反序列化，不仅取决于类路径和功能代码是否一致，还有非常重要的一点是两个类的序列化 ID 是否一致（一般声明为 private static final long serialVersionUID = 1L）。

3. Android 中的序列化

Android 序列化对象主要有两种方法，实现 Serializable 接口或者实现 Parcelable 接口。实现 Serializable 接口是 Java SE 本身就支持的，而 Parcelable 是 Android 特有的功能，效率比实现 Serializable 接口高，而且还可以用在 IPC 中。实现 Serializable 接口非常简单，声明一下就可以了；而实现 Parcelable 接口稍微复杂一些，但效率更高，推荐用这种方法提高性能。

通过实现 Parcelable 接口序列化对象的基本步骤如下：

（1）声明实现接口 Parcelable。

（2）实现 Parcelable 的 writeToParcel()方法，将对象序列化为一个 Parcel 对象。

（3）实例化静态内部对象 CREATOR 实现接口 Parcelable. Creator，内部对象 CREATOR 的名称必须全部大写。

（4）完成 CREATOR 的功能，实现 createFromParcel()方法，将 Parcel 对象反序列化为目标对象。

下面的示例演示了 Android 中 Activity 之间传递实现 Parcelable 接口序列化信息的方法。

```
public class TransmitComplexDataActivity extends Activity {
    private Button mButton;

    @Override
    public void onCreate(Bundle savedInstanceState) {
        super.onCreate(savedInstanceState);
        setContentView(R.layout.activity_transmit_complex_data);

        mButton = (Button) this.findViewById(R.id.button1);
        mButton.setOnClickListener(new OnClickListener() {

            @Override
            public void onClick(View v) {
                startTransmitParcelableData();
            }

        });
    }

    private void startTransmitParcelableData() {
```

```
        Book mBook = new Book();
        mBook.setBookName("Android Tutor");
        mBook.setAuthor("Liweiyong");
        Intent mIntent = new Intent(this, ShowComplexDataActivity.class);
        Bundle mBundle = new Bundle();
        mBundle.putParcelable("parcelableData", mBook);
        mBundle.putInt("type", 2);
        mIntent.putExtras(mBundle);
        startActivity(mIntent);
    }

}
```

其中,Book 类实现了 Parcelable 接口,代码如下:

```
public class Book implements Parcelable {
    private String bookName;
    private String author;

    public String getBookName() {
        return bookName;
    }

    public void setBookName(String bookName) {
        this.bookName = bookName;
    }

    public String getAuthor() {
        return author;
    }

    public void setAuthor(String author) {
        this.author = author;
    }

    public static final Parcelable.Creator<Book> CREATOR
            = new Creator<Book>() {
        public Book createFromParcel(Parcel source) {
            Book mBook = new Book();
            mBook.bookName = source.readString();
            mBook.author = source.readString();
            return mBook;
        }

        public Book[] newArray(int size) {
            return new Book[size];
        }
    };

```

```
    public int describeContents() {
        return 0;
    }

    public void writeToParcel(Parcel parcel, int flags) {
        parcel.writeString(bookName);
        parcel.writeString(author);
    }
}
```

1.4.2　编码转换

现实世界存在着多种语言，表示这些语言的符号也千差万别，而计算机中存储信息的最小单元是一个字节(8 个 bit)，能够表示的字符范围是 0～255 个，这远远不能满足人类表示语言符号的要求。因此，两台交互的计算机之间经常需要对交互信息进行编码转换，以适应人类的交互。

1. 常见编码

计算机提供的常见编码方式包括以下几种。

(1) ASCII 码

标准 ASCII 码用一个字节的低 7 位表示符号，能够表示 128 种字符。其中，0～31 是控制字符(如换行、回车等)，32～126 是打印字符，可以通过键盘输入并且能够显示出来。

(2) ISO-8859-1

标准 ASCII 码只能表示 128 个字符，这显然是不够用的。于是 ISO 组织在 ASCII 码基础上又制定了一系列标准用来扩展 ASCII 编码，它们是 ISO-8859-1～ISO-8859-15，其中 ISO-8859-1 涵盖了大多数西欧语言字符，应用也最广泛。ISO-8859-1 仍然是单字节编码，它总共能表示 256 个字符。

(3) GB2312/GBK/GB18030

GB2312 的全称是《信息交换用汉字编码字符集　基本集》，它是双字节编码，总的编码范围是 A1～F7，其中 A1～A9 是符号区，总共包含 682 个符号；B0～F7 是汉字区，包含 6763 个汉字。

GBK 的全称是《汉字内码扩展规范》，它扩展 GB2312，加入了更多的汉字，它的编码范围是 8140～FEFE(去掉 XX7F)，总共有 23940 个码位，它能表示 21003 个汉字，它的编码和 GB2312 兼容。

GB18030 的全称是《信息交换用汉字编码字符集》，是我国的强制标准，它可能是单字节、双字节或者四字节编码，它的编码与 GB2312 编码兼容，这个虽然是国家标准，但是实际应用系统中使用得并不广泛。

(4) UTF-16

UTF-16 具体定义了 Unicode 字符在计算机中的存取方法。UTF-16 用两个字节来表示 Unicode 转化格式，这个是定长的表示方法，不论什么字符都可以用两个字节表示。UTF-16 表示字符非常方便，每两个字节表示一个字符，这样就大大简化了字符串的操作，这也是 Java 以 UTF-16 作为内存的字符存储格式一个很重要的原因。

(5) UTF-8

UTF-16 统一采用两个字节表示一个字符,虽然在表示上非常简单方便,但是也有其缺点,有很多字符用一个字节就可以表示,现在要用两个字节表示,存储空间加大了一倍,这样会增大网络传输的流量,而且也没必要。而 UTF-8 采用了一种变长技术,每个编码区域有不同的字码长度。不同类型的字符可以由 1～6 个字节组成。

(6) UTF-8 有以下编码规则:

① 如果一个字节最高位(第 8 位)为 0,表示这是一个 ASCII 字符(00～7F)。因此,所有 ASCII 编码已经是 UTF-8 了。

② 如果一个字节以 11 开头,连续的 1 的个数表示这个字符的字节数,例如: 110×××××代表它是双字节 UTF-8 字符的首字节。

③ 如果一个字节以 10 开始,表示它不是首字节,需要向前查找才能得到当前字符的首字节。

2. 常见编码转换

(1) FileReader 转为 BufferedReader:

```
BufferedReader in = new BufferedReader(new FileReader("filename.java"));
```

(2) InputStream 转为 BufferedReader:

```
BufferedReader in = new BufferedReader(new InputStreamReader(System.in));
```

(3) String 转为 DataInputStream:

```
DataInputStream in = new DataInputStream(new ByteArrayInputStream(str.getBytes()));
```

(4) FileInputStream 转为 DataInputStream:

```
DataInputStream in = new DataInputStream ( new BufferedInputStream ( new FileInputStream
("filename.txt")));
```

(5) FileWriter 转为 PrintWriter:

```
PrintWriter pw = new PrintWriter(new BufferedWriter("filename.out"));
```

(6) FileOutputStream 转为 PrintStream:

```
PrintStream ps = new PrintStream ( new BufferedOutputStream ( new FileOutputStream ( " text.
out")));
```

(7) FileOutputStream 转为 DataOutputStream：

```
DataOutputStream dos = new DataOutputStream(new BufferedOutputStream(new FileOutputStream("
filename.txt")));
```

1.5 习 题

1. 编程实现对应用中图片的缓存处理。
2. 编程实现将一个应用的登录信息通过序列化方式在 Activity 之间共享。

第2章 数据解析

当移动客户端向移动互联网发送请求信息时，客户端与服务端之间的数据交互通常以XML或JSON格式响应，这种格式为丰富移动互联网应用提供了巨大的支持。

2.1 XML数据解析

Java平台支持通过许多不同的方式来使用XML，并且大多数与XML相关的Java API在Android上得到了完全支持。

2.1.1 XML介绍

XML(Extensible Markup Language，可扩展标记语言)是一种易于使用和易于扩展的标记语言，是SGML(The Standard Generalized Markup Language，标准通用标记语言)的子集，主要用于Internet的跨平台数据传递。

1. XML的结构

XML文档是由一组使用唯一名称标识的实体组成的，XML是一种树结构，它从“根部”开始，然后扩展到“枝叶”。XML文档由内容和标记(包括声明、元素、注释、字符引用等)组成，通过以标记包围内容的方式将大部分内容包含在元素中。下面是一个简单的XML文档示例：

```
01 <?xml version = "1.0" standalone = "yes" encoding = "ISO - 8859 - 1"?>
02 <! -- Copyright w3school.com.cn -->
03 < note >
04     < to > George </to >
05     < from > John </from >
06     < heading > Reminder </heading >
07     < body > Don't forget the meeting!</body >
08 </note >
```

01行是一个XML处理指令。处理指令以“＜?”开始，以“? ＞”结束。“＜?”后的第一个单词是指令名，如xml，代表XML声明。version、standalone、encoding是三个特性，特性是由等号分开的名称—数值对，等号左边是特性名称，等号右边是特性的值，用引号引起来。version＝"1.0"说明这个文档符合1.0规范；standalone说明文档在这个文件里还是需要从外部导入，standalone＝"yes"说明所有的文档都在这一文件里完成；encoding指文档字符编码。

02 行是对 XML 的注释说明。需要注意的是：注释中不要出现"——"或"—"；注释不要放在标记中；注释不能嵌套。

03～08 行是对 XML 根元素的定义和描述。XML 文档的树形结构要求必须有一个根元素。根元素的起始标记要放在所有其他元素起始标记之前，根元素的结束标记要放在其他所有元素的结束标记之后。

04、05、06、07 行是 4 个 XML 元素。元素的基本结构由"开始标记，数据内容，结束标记"组成。其中，数据内容中的空格不会被删除或合并。需要注意的是：元素标记区分大小写，<From>与<from>是两个不同的标记；结束标记必须有反斜杠，如</from>。

XML 元素标记命名规则如下：

- 名字中可以包含字母、数字及其他规定的字符。
- 名字不能以数字或下划线开头。
- 名字不能用 xml 开头。
- 名字中不能包含空格和冒号。

2. XML 的语法

(1) 所有 XML 元素都必须有关闭标签

在 XML 中，省略关闭标签是非法的。所有元素都必须有关闭标签，如：

```
<p>This is a paragraph</p>
```

注意：XML 声明没有关闭标签。这是因为声明不属于 XML 本身的组成部分。它不是 XML 元素，也不需要关闭标签。

(2) XML 标签对大小写敏感

XML 元素使用 XML 标签进行定义。XML 标签对大小写敏感。必须使用相同的大小写来编写打开标签和关闭标签(打开标签和关闭标签通常被称为开始标签和结束标签)，如：

```
<Message>这是错误的</message>
<message>这是正确的</message>
```

(3) XML 必须正确地嵌套

在 XML 中，所有元素都必须彼此正确地嵌套，如下例中<i>元素是在<b>元素内打开的，那么它必须在<b>元素内关闭(注意，包含了子元素标签的标签一般不包含数据值)：

```
<b><i>This text is bold and italic</i></b>
```

(4) XML 的属性值必须加引号

在 XML 中，有时候要为元素添加属性。属性由一个键值对构成，XML 的属性值必须加引号。下面的 XML 文档是错误的：

```
<note date=08/08/2008>
    <to>George</to>
    <from>John</from>
</note>
```

属性是在使用元素时存储额外信息的一种方式。在同一个文档中,可以根据需要对每个元素的不同实例采用不同的属性值。

(5) 实体引用

不能直接在 XML 文档内容中输入特殊字符。如果要在文本中使用符号,必须使用它的字符代码,即实体。可以将短语(例如公司名)设置为实体,然后就可以在内容中使用该实体。为了设置实体,必须先为它创建一个名称,然后将它输入内容中。实体以 & 符号开始,并以分号结束,例如"&coname;"。然后在 DOCTYPE 的方括号内部输入代码,这个代码识别表示实体的文本。

在 XML 中,有 5 个预定义的实体引用,如表 2-1 所示。

表 2-1　预定义实体引用

实体引用	符　号	含　义
<	<	小于
>	>	大于
&	&	和号
'	'	单引号
"	"	双引号

例如,下例中,01 行是错误的,02 行是正确的。

```
01 <message> if salary < 1000 then </message>
02 <message> if salary &lt; 1000 then </message>
```

必须正确地声明和表示实体,以避免错误和确保正确显示。

(6) XML 的元素可以循环嵌套

嵌套即把某个元素放到其他元素的内部。这些新的元素称为子元素,包含它们的元素称为父元素。父级元素包含子级元素,子级元素又可以包含自己的子级元素,例如:

```
<?xml version = "1.0" encoding = "ISO - 8859 - 1"?>
<! -- Edited with XML Spy v2007 (http://www.altova.com) -->
<breakfast_menu>
    <food>
        <name> Belgian Waffles </name>
        <price> $ 5.95 </price>
        <description> two of our famous Belgian Waffles with plenty of
                real maple syrup </description>
        <calories> 650 </calories>
    </food>
    <food>
        <name> Strawberry Belgian Waffles </name>
        <price> $ 7.95 </price>
        <description> light Belgian waffles covered with strawberries
                and whipped cream </description>
        <calories> 900 </calories>
```

```
    </food>
    <food>
        <name>Berry-Berry Belgian Waffles</name>
        <price>$8.95</price>
        <description>light Belgian waffles covered with an assortment of
                fresh berries and whipped cream</description>
        <calories>900</calories>
    </food>
    <food>
        <name>French Toast</name>
        <price>$4.50</price>
        <description>thick slices made from our homemade sourdough
                bread</description>
        <calories>600</calories>
    </food>
    <food>
        <name>Homestyle Breakfast</name>
        <price>$6.95</price>
        <description>two eggs, bacon or sausage, toast, and our
                ever-popular hash browns</description>
        <calories>950</calories>
    </food>
</breakfast_menu>
```

一个常见的语法错误是父元素和子元素的错误嵌套。任何子元素都要完全包含在其父元素的开始和结束标记内部。每个同胞元素必须在下一个同胞元素开始之前结束。

3. XML 的验证

构造良好的 XML 即遵循所有 XML 规则创建的 XML：正确的元素命名、嵌套、属性命名等。

验证就是根据元素规则检查文档的结构，以及如何为每个父元素定义子元素。这些规则是在文档类型定义(Document Type Definition，DTD)或模式(Schema)中定义的。

(1) DTD 验证

DTD 规定了文档的逻辑结构。它可定义文档的语法，而文档的语法反过来也能够让 XML 语法分析程序确认页面标记使用的合法性。DTD 定义了页面的元素、元素的属性及元素和属性间的关系。元素与元素间用起始标记和结束标记来定界。对于空元素，用一个空元素标记来分隔。每一个元素都有一个用名字标识的类型，也称为它的通用标识符，并且它还可以有一个属性说明集。每个属性说明都有一个名字和一个值。理想定义应该面向描述与应用程序相关的数据结构，而不是如何显示数据。也就是说，应该把一个元素定义为一个标题行，之后让样式表和脚本定义显示标题行。

合法的 XML 文档是遵守文档类型定义语法规则的形式良好的 XML 文档。DTD 分为以下三类。

- 内部 DTD 文档

```
<!DOCTYPE 根元素 [定义内容]>
```

• 外部 DTD 文档

```
<!DOCTYPE 根元素 SYSTEM "DTD 文件路径">
```

• 内外部 DTD 文档结合

```
<!DOCTYPE 根元素 SYSTEM "DTD 文件路径" [定义内容]>
```

例如：

```
<?xml version = "1.0" encoding = "ISO - 8859 - 1"?>
<!DOCTYPE note SYSTEM "Note.dtd">
< note >
    < to > George </to >
    < from > John </from >
    < heading > Reminder </heading >
    < body > Don't forget the meeting! </body >
</note >
```

在上例中,DOCTYPE 声明是对外部 DTD 文件的引用。DTD 的作用是定义 XML 文档的结构。它使用一系列合法的元素来定义文档结构,例如：

```
<!DOCTYPE note [
    <!ELEMENT note (to,from,heading,body)>
    <!ELEMENT to      (#PCDATA)>
    <!ELEMENT from    (#PCDATA)>
    <!ELEMENT heading (#PCDATA)>
    <!ELEMENT body    (#PCDATA)>
]>
```

DTD 不具有强制性。对于简单的应用程序来说,开发商不需建立自己的 DTD,可以使用预先定义的公共 DTD 或不使用。即使某个文档已经有 DTD,只要文档组织是良好的,语法分析程序也不必对照 DTD 来检验文档的合法性。服务器可能已执行了检查,所以检验的时间和带宽将得以大幅度节省。

(2) XML Schema 验证

XML Schema 是一组用于约束结构和清晰表达 XML 文档的信息集规则。XML Schema 是用一套预先规定的 XML 元素和属性创建的,这些元素和属性定义了 XML 文档的结构和内容模式。XML Schema 规定 XML 文档实例的结构和每个元素/属性的数据类型。例如：

```
< xs:element name = "note">
    < xs:complexType >
        < xs:sequence >
            < xs:element name = "to"      type = "xs:string"/>
```

```
                <xs:element name = "from"     type = "xs:string"/>
                <xs:element name = "heading" type = "xs:string"/>
                <xs:element name = "body"     type = "xs:string"/>
            </xs:sequence>
        </xs:complexType>

</xs:element>
```

所有 Schema 文档使用 schema 作为其根元素，用于构造 schema 的元素和数据类型来自 http://www.w3.org/2001/XMLSchema 命名空间。

Schema 的特性包括以下方面。

- Schema 基于 XML 语法。
- Schema 可以用能处理 XML 文档的工具处理。
- Schema 大大扩充了数据类型，可以自定义数据类型。
- Schema 支持元素的继承——Object-Oriented。
- Schema 支持属性组。

XML Schema 的作用包括：

- 定义可出现在文档中的元素。
- 定义可出现在文档中的属性。
- 定义哪个元素是子元素。
- 定义子元素的次序。
- 定义子元素的数目。
- 定义元素是否为空，或者是否可包含文本。
- 定义元素和属性的数据类型。
- 定义元素和属性的默认值以及固定值。

2.1.2　Android 中的 XML 解析

在 Android 中，常见的 XML 解析器分别为 DOM 解析、SAX 解析和 PULL 解析。

1. DOM 解析

DOM(Document Object Model)是 W3C 处理 XML 的标准 API，允许开发人员使用 DOM API 遍历 XML 树、检索所需数据。DOM 是基于树形结构的节点或信息片段的集合，分析该结构通常需要加载整个文档和构造树形结构，然后才可以检索和更新节点信息。

DOM 解析的优点是：由于 DOM 的处理方式是将 XML 整个作为类似树结构的方式读入内存中，因此检索和更新效率会更高，同时支持应用程序对 XML 数据进行删除、修改、重新排列等多种操作。但是由于其需要在处理开始时将整个 XML 文件读入内存中去进行分析，因此其在解析大数据量的 XML 文件时会遇到类似于内存泄露以及程序崩溃的风险。因此，DOM 解析方式适合于小型 XML 文件解析、需要全解析或者大部分解析 XML、需要修改 XML 树内容以生成自己的对象模型。

DOM 解析的基本步骤是：

① 将 XML 文件加载进来；

② 获取文档的根节点；

③ 获取文档根节点中所有子节点的列表；

④ 获取子节点列表中需要读取的节点信息。

例如，有下面的 XML 文档 books. xml：

```
<?xml version="1.0" encoding="utf-8"?>
<books>
<book>
        <id>1001</id>
        <name>Thinking In Java</name>
        <price>80.00</price>
    </book>
    <book>
        <id>1002</id>
        <name>Core Java</name>
        <price>90.00</price>
    </book>
    <book>
        <id>1003</id>
        <name>Hello, Andriod</name>
        <price>100.00</price>
    </book>
</books>
```

在 Activity 的 parsingXML()方法中调用如下代码：

```
try {
    InputStream is = getAssets().open("books.xml");
    parser = new DomBookParser();
    books = parser.parse(is);
    for (Book book : books) {
        Log.i(TAG, book.toString());
    }
    } catch (Exception e) {
    Log.e(TAG, e.getMessage());
}
```

其中的 parse()方法代码如下：

```
@Override
public List<Book> parse(InputStream is) throws Exception {
    List<Book> books = new ArrayList<Book>();
    DocumentBuilderFactory factory = DocumentBuilderFactory.newInstance();
    DocumentBuilder builder = factory.newDocumentBuilder();
    Document doc = builder.parse(is);
```

```
    Element rootElement = doc.getDocumentElement();
    NodeList items = rootElement.getElementsByTagName("book");

    for (int i = 0; i < items.getLength(); i++) {
        Book book = new Book();
        Node item = items.item(i);
        NodeList properties = item.getChildNodes();

        for (int j = 0; j < properties.getLength(); j++) {
            Node property = properties.item(j);
            String nodeName = property.getNodeName();

            if (nodeName.equals("id")) {
                book.setId(Integer.parseInt(property.getFirstChild()
                        .getNodeValue()));
            } else if (nodeName.equals("name")) {
                book.setName(property.getFirstChild().getNodeValue());
            } else if (nodeName.equals("price")) {
                book.setPrice(Float.parseFloat(property.getFirstChild()
                        .getNodeValue()));
            }
        }

        books.add(book);
    }

    return books;
}
```

1）javax.xml.parsers 包的 API 简介

（1）DocumentBuilderFactory

定义工厂 API，使应用程序能够从 XML 文档获取生成 DOM 对象树的解析器。在这里使用 DocumentBuilderFacotry 是为了创建与具体解析器无关的程序，当 DocumentBuilderFactory 类的静态方法 newInstance()被调用时，它根据一个系统变量来决定具体使用哪一个解析器。又因为所有的解析器都服从于 JAXP 所定义的接口，所以无论具体使用哪一个解析器，代码都是一样的。所以当在不同的解析器之间进行切换时，只需要更改系统变量的值，而不用更改任何代码。这就是工厂所带来的好处。

- DocumentBuilderFactory 提供的重要方法包括以下几种。
- newInstance()：获取 DocumentBuilderFactory 的新实例。

newDocumentBuilder()：使用当前配置的参数创建一个新的 DocumentBuilder 实例。

（2）DocumentBuilder

定义 API，使其从 XML 文档获取 DOM 文档实例。当获得一个 DocumentBuilderFactory 工厂对象后，使用它的静态方法 newDocumentBuilder()可以获得一个 DocumentBuilder 对象，这个对象代表了具体的 DOM 解析器。

DocumentBuilder 提供的重要方法包括以下几种。

- parse(File f)：将给定文件的内容解析为一个 XML 文档，并且返回一个新的 DOM Document 对象。
- parse(InputSource is)：将给定输入源的内容解析为一个 XML 文档，并且返回一个新的 DOM Document 对象。
- parse(InputStream is)：将给定 InputStream 的内容解析为一个 XML 文档，并且返回一个新的 DOM Document 对象。
- parse(InputStream is, String systemId)：将给定 InputStream 的内容解析为一个 XML 文档，并且返回一个新的 DOM Document 对象。
- parse(String uri)：将给定 URI 的内容解析为一个 XML 文档，并且返回一个新的 DOM Document 对象。
- newDocument()：获取 DOM Document 对象的一个新实例来生成一个 DOM 树。

2) org. w3c. dom 包的 API 简介

(1) Document

Document 对象代表了整个 XML 的文档，所有其他的 Node 都以一定的顺序包含在 Document 对象内，排列成一个树形的结构，程序员可以通过遍历这棵树来得到 XML 文档的所有内容，这也是对 XML 文档操作的起点。

Document 接口提供的重要方法包括以下几种。

- createAttribute(String)：用给定的属性名创建一个 Attr 对象，并可在其后使用 setAttributeNode()方法来放置在某一个 Element 对象上面。
- createElement(String)：用给定的标签名创建一个 Element 对象，代表 XML 文档中的一个标签，然后就可以在这个 Element 对象上添加属性或进行其他的操作。
- createTextNode(String)：用给定的字符串创建一个 Text 对象，Text 对象代表了标签或者属性中所包含的纯文本字符串。如果在一个标签内没有其他标签，那么标签内的文本所代表的 Text 对象就是这个 Element 对象的唯一子对象。
- getDocumentElement()：返回一个代表这个 DOM 树的根节点的 Element 对象，也就是代表 XML 文档根元素的那个对象。
- getDocumentURI()：文档的位置，如果未定义或 Document 是使用 DOMImplementation. createDocument 创建的，则为 null。
- getDomConfig()：调用 Document. normalizeDocument()时使用的配置。
- getElementById(String elementId)：返回具有带给定值的 ID 属性的 Element。
- getElementsByTagName(String tagname)：按文档顺序返回包含在文档中且具有给定标记名称的所有 Element 的 NodeList。
- getElementsByTagNameNS(String namespaceURI, String localName)：以文档顺序返回具有给定本地名称和名称空间 URI 的所有 Elements 的 NodeList。
- getImplementation()：处理此文档的 DOMImplementation 对象。
- getXmlEncoding()：作为 XML 声明的一部分，指定此文档编码的属性。
- getXmlStandalone()：作为 XML 声明的一部分，指定此文档是否为独立文档的属性。

- getXmlVersion()：作为 XML 声明的一部分，指定此文档版本号的属性。
- importNode(Node importedNode, boolean deep)：从另一文档向此文档导入节点，而不改变或移除原始文档中的源节点。此方法创建源节点的一个新副本。
- normalizeDocument()：此方法的行为如同使文档通过一个保存和加载的过程，而将其置为 normal(标准)形式。
- renameNode(Node n, String namespaceURI, String qualifiedName)：重命名 ELEMENT_NODE 或 ATTRIBUTE_NODE 类型的现有节点。
- normalize：可以去掉 XML 文档中作为格式化内容的空白而映射在 DOM 树中的不必要的 Text Node 对象。

(2) Node

Node 对象是 DOM 结构中最为基本的对象，代表了文档树中的一个抽象的节点。在实际使用的时候，很少会真正用到 Node 这个对象，而是用到诸如 Element、Attr、Text 等 Node 对象的子对象来操作文档。Node 对象为这些对象提供了一个抽象的、公共的根。虽然在 Node 对象中定义了对其子节点进行存取的方法，但是有一些 Node 子对象，例如 Text 对象，它并不存在子节点，这一点是要注意的。

Node 接口提供的重要方法包括以下几种。

- appendChild(org.w3c.dom.Node)：为这个节点添加一个子节点，并放在所有子节点的最后。如果这个子节点已经存在，则先把它删掉再添加进去。
- getFirstChild()：如果节点存在子节点，则返回第一个子节点。对等的，还有 getLastChild()方法返回最后一个子节点。
- getNextSibling()：返回在 DOM 树中这个节点的下一个兄弟节点。对等的，还有 getPreviousSibling()方法返回其前一个兄弟节点。
- getNodeName()：根据节点的类型返回节点的名称。
- getNodeType()：返回节点的类型。
- getNodeValue()：返回节点的值。
- hasChildNodes()：判断是不是存在子节点。
- hasAttributes()：判断这个节点是否存在属性。
- getOwnerDocument()：返回节点所处的 Document 对象。
- insertBefore(org.w3c.dom.Node new,org.w3c.dom.Node ref)：在给定的一个子对象前再插入一个子对象。
- removeChild(org.w3c.dom.Node)：删除给定的子节点对象。
- replaceChild(org.w3c.dom.Node new,org.w3c.dom.Node old)：用一个新的 Node 对象代替给定的子节点对象。

(3) NodeList

NodeList 对象代表了一个包含一个或者多个 Node 的列表。可以简单地把它看成一个 Node 的数组。

NodeList 接口提供的重要方法包括以下几种。

- GetLength()：返回列表的长度。
- Item(int)：返回指定位置的 Node 对象。

(4) Element

Element 对象代表的是 XML 文档中的标签元素,继承自 Node,亦是 Node 的最主要的子对象。在标签中可以包含属性,因而 Element 对象中有存取其属性的方法,而任何 Node 中定义的方法也可以用在 Element 对象上面。

Element 接口提供的重要方法包括以下几种。

- getElementsByTagName(String):返回一个 NodeList 对象,它包含了在这个标签之下的子孙节点中具有给定标签名字的标签。
- getTagName():返回一个代表这个标签名字的字符串。
- getAttribute(String):返回标签中给定属性名称的属性值。需要注意的是,XML 文档中应允许有实体属性出现,而这个方法对这些实体属性并不适用。这时候需要用 getAttributeNodes()方法来得到一个 Attr 对象进行进一步的操作。
- getAttributeNode(String):返回一个代表给定属性名称的 Attr 对象。

(5) Attr

Attr 对象代表了某个标签中的属性。Attr 继承自 Node,但是因为 Attr 实际上是包含在 Element 中的,它并不能被看作 Element 的子对象,因而在 DOM 中 Attr 并不是 DOM 树的一部分,所以 Node 中的 getParentNode()、getPreviousSibling()和 getNextSibling()返回的都将是 null。也就是说,Attr 其实是被看作包含它的 Element 对象的一部分,它并不作为 DOM 树中单独的一个节点出现。这一点在使用的时候要同其他 Node 子对象相区别。

Attr 接口提供的重要方法包括以下几种。

- getName():返回此属性的名称。
- getOwnerElement():此属性连接到的 Element 节点。如果未使用此属性,则为 null。
- getSchemaTypeInfo():与此属性相关联的类型信息。
- getSpecified():如果在实例文档中显式给此属性一个值,则为 True;否则为 False。
- getValue():检索时,该属性值以字符串形式返回。
- isId():返回此属性是否属于类型 ID(即包含其所有者元素的标识符)。
- setValue(String value):检索时,该属性值以字符串形式返回。

下面的 serialize()方法演示了将 List 集合数据生成 XML 文件的方法。

```
@Override
public String serialize(List<Book> books) throws Exception {
    DocumentBuilderFactory factory = DocumentBuilderFactory.newInstance();
    DocumentBuilder builder = factory.newDocumentBuilder();
    Document doc = builder.newDocument();

    Element rootElement = doc.createElement("books");

    for (Book book : books) {
        Element bookElement = doc.createElement("book");
        bookElement.setAttribute("id", book.getId() + "");

```

```
        Element nameElement = doc.createElement("name");
        nameElement.setTextContent(book.getName());
        bookElement.appendChild(nameElement);

        Element priceElement = doc.createElement("price");
        priceElement.setTextContent(book.getPrice() + "");
        bookElement.appendChild(priceElement);

        rootElement.appendChild(bookElement);
    }

    doc.appendChild(rootElement);

    TransformerFactory transFactory = TransformerFactory.newInstance();
    Transformer transformer = transFactory.newTransformer();
    transformer.setOutputProperty(OutputKeys.ENCODING, "UTF-8");
    transformer.setOutputProperty(OutputKeys.INDENT, "yes");
    transformer.setOutputProperty(OutputKeys.OMIT_XML_DECLARATION, "no");

    StringWriter writer = new StringWriter();

    Source source = new DOMSource(doc);
    Result result = new StreamResult(writer);
    transformer.transform(source, result);

    return writer.toString();
}
```

2. SAX 解析

SAX(Simple API for XML)解析器是一种基于事件驱动的解析器。所谓事件驱动，就是它不用解析完整个文档，在按内容顺序解析文档过程中，SAX 会判断当前读到的字符是否符合 XML 文件语法中的某部分。如果符合某部分，则会触发事件。所谓触发事件，就是调用一些回调方法，用 SAX 解析 XML 文档时，在读取到文档开始和结束标签时就会回调一个事件，在读取到其他节点与内容时也会回调一个事件。在事件源调用事件处理器中特定方法的时候，还要传递给事件处理器相应事件的状态信息，这样事件处理器才能够根据提供的事件信息来决定自己的行为。

SAX 从根本上解决了 DOM 在解析 XML 文档时产生的占用大量资源的问题。其实现是通过类似于流解析的技术，通读整个 XML 文档树，通过事件处理器来响应程序员对于 XML 数据解析的需求。由于其不需要将整个 XML 文档读入内存中，它对系统资源的节省是显而易见的。SAX 解析方式适合于大型 XML 文件解析、只需要部分解析或者只想取得部分 XML 树内容、有 XPath 查询需求、有自己生成特定 XML 树对象模型需求的情况。

SAX 支持 XPath 查询，使得开发人员处理 XML 更加灵活和得心应手。但是其仍然有一些不足之处困扰着广大的开发人员：首先，它十分复杂的 API 接口令人望而生畏；其次，由于其是类似流解析的文件扫描方式，因此不支持应用程序对 XML 树内容结构等的修改，可能会有不便之处。

SAX 解析的基本步骤：

① 创建一个 SAXParserFactory 对象。

② 调用 SAXParserFactory 中的 newSAXParser 方法创建一个 SAXParser 对象。

③ 然后调用 SAXParser 中的 getXMLReader 方法获取一个 XMLReader 对象。

④ 实例化一个 DefaultHandler 对象。

⑤ 连接事件源对象 XMLReader 到事件处理类 DefaultHandler 中。

⑥ 调用 XMLReader 的 parse 方法从输入源中获取的 XML 数据。

⑦ 通过 DefaultHandler 返回需要的数据集合。

例如，解析上例中的 XML 文档 books. xml，parse()方法代码如下：

```
@Override
public List<Book> parse(InputStream is) throws Exception {
    SAXParserFactory factory = SAXParserFactory.newInstance();
    SAXParser parser = factory.newSAXParser();
    XMLContentHandler handler = new XMLContentHandler();
    parser.parse(is, handler);

    return handler.getBooks();
}
```

其中的 XMLContentHandler 代码如下：

```
class XMLContentHandler extends DefaultHandler {

    private List<Book> books;
    private Book book;
    private StringBuilder builder;

    public List<Book> getBooks() {
        return books;
    }

    @Override
    public void startDocument() throws SAXException {
        super.startDocument();
        books = new ArrayList<Book>();
        builder = new StringBuilder();
    }

    @Override
    public void startElement(String uri, String localName, String qName,
            Attributes attributes) throws SAXException {
        super.startElement(uri, localName, qName, attributes);
        if (localName.equals("book")) {
            book = new Book();
```

```
        }
        builder.setLength(0);
    }

    @Override
    public void characters(char[] ch, int start, int length)
            throws SAXException {
        super.characters(ch, start, length);
        builder.append(ch, start, length);
    }

    @Override
    public void endElement(String uri, String localName, String qName)
            throws SAXException {
        super.endElement(uri, localName, qName);
        if (localName.equals("id")) {
            book.setId(Integer.parseInt(builder.toString()));
        } else if (localName.equals("name")) {
            book.setName(builder.toString());
        } else if (localName.equals("price")) {
            book.setPrice(Float.parseFloat(builder.toString()));
        } else if (localName.equals("book")) {
            books.add(book);
        }
    }
}
```

下面介绍 org. xml. sax 包的 API。

在 SAX 接口中，事件源是 org. xml. sax 包中的 XMLReader，它通过 parser()方法来解析 XML 文档，并产生事件。事件处理器是 org. xml. sax 包中 ContentHander、DTDHander、ErrorHandler 以及 EntityResolver 这 4 个接口。

DefaultHandler 是一个事件处理器，可以接收解析器报告的所有事件，处理所发现的数据。它实现了 ContentHandler 接口、DTDHandler 接口、ErrorHandler 接口和 EntityResolver 接口。这几个接口代表不同类型的事件处理器。这 4 个接口的详细说明如表 2-2 所示。

表 2-2　org. xml. sax 包中的事件处理器

事件处理器名称	事件处理器处理的事件	XMLReader 注册方法
ContentHander	跟文档内容有关的所有事件： (1) 文档的开始和结束。 (2) XML 元素的开始和结束。 (3) 可忽略的实体。 (4) 名称空间前缀映射的开始和结束。 (5) 处理指令。 (6) 字符数据和可忽略的空格	setContentHandler(ContentHandler h)

续表

事件处理器名称	事件处理器处理的事件	XMLReader 注册方法
DTDHander	处理对文档 DTD 进行解析时产生的相应事件	setDTDHandler(DTDHandler h)
ErrorHandler	处理 XML 文档解析时产生的错误。如果一个应用程序没有注册一个错误处理器类,会发生不可预料的解析器行为	setErrorHandler(ErrorHandler h)
EntityResolver	处理外部实体	setEntityResolver(EntityResolver e)

ContentHandler 接口常用的方法包括以下几种。

- startDocument():当遇到文档开头的时候,调用这个方法,可以在其中做一些预处理的工作。
- endDocument():当文档结束的时候,调用这个方法,可以在其中做一些善后的工作。
- startElement(String namespaceURI, String localName,String qName, Attributes atts):当读到开始标签的时候,会调用这个方法。namespaceURI 就是命名空间,localName 是不带命名空间前缀的标签名,qName 是带命名空间前缀的标签名。通过 atts 可以得到所有的属性名和相应的值。
- endElement(String uri, String localName, String name):在遇到结束标签的时候,调用这个方法。
- characters(char[] ch, int start, int length):这个方法用来处理在 XML 文件中读到的内容。例如:<high data="30"/>主要目的是获取 high 标签中的值。第一个参数用于存放文件的内容,后面两个参数是读到的字符串在这个数组中的起始位置和长度,使用 new String(ch,start,length)就可以获取内容。

注意:SAX 的一个重要特点就是它的流式处理,当遇到一个标签的时候,它并不会记录下之前所碰到的标签,即在 startElement()方法中,所有能够知道的信息,就是标签的名字和属性。至于标签的嵌套结构、上层标签的名字、是否有子元素等其他与结构相关的信息,都是不知道的,都需要程序来完成。这使得 SAX 在编程处理上没有 DOM 方便。

表 2-3 列出了 SAX 和 DOM 在一些方面的对照。

表 2-3 SAX 和 DOM 对比

SAX	DOM
顺序读入文档并产生相应事件,可以处理任何大小的 XML 文档	在内存中创建文档树,不适于处理大型 XML 文档
只能对文档按顺序解析一遍,不支持对文档的随意访问	可以随意访问文档树的任何部分,没有次数限制
只能读取 XML 文档内容,而不能修改	可以随意修改文档树,从而修改 XML 文档
开发上比较复杂,需要自己来实现事件处理器	易于理解,易于开发
对开发人员而言更灵活,可以用 SAX 创建自己的 XML 对象模型	已经在 DOM 基础之上创建好了文档树

3. PULL 解析

PULL 解析器的运行方式和 SAX 类似，都是基于事件的模式。不同的是，在 PULL 解析过程中，需要自己获取产生的事件然后做相应的操作，而不像 SAX 那样由处理器触发一种事件的方法来执行相应的工作。PULL 解析器小巧轻便，解析速度快，简单易用，非常适合在 Android 移动设备中使用，Android 系统内部在解析各种 XML 时也是用 PULL 解析器。

Pull 解析是一个遍历文档的过程，每次调用 next()、nextTag()、nextToken()和 nextText()都会向前推进文档，并使 Parser 停留在某些事件上面，但是不能倒退。然后把文档设置给 Parser。

Android 中对 Pull 方法提供了支持的 API，常用的 XML PULL 的接口和类包括以下几种。

- XmlPullParser：该解析器是一个在 org. xmlpull. v1 中定义的解析功能的接口。应用程序通过调用 XmlPullParser. next()等方法来产生 Event，然后再处理 Event。
- XmlSerializer：它是一个接口，定义了 XML 信息集的序列。
- XmlPullParserFactory：这个类用于在 XMPULL V1 API 中创建 XML Pull 解析器。
- XmlPullParserException：抛出单一的与 XML pull 解析器相关的错误。

例如，解析上例中的 XML 文档 books. xml，parse()方法代码如下：

```
@Override
public List<Book> parse(InputStream is) throws Exception {
    List<Book> books = null;
    Book book = null;

    XmlPullParser parser = Xml.newPullParser();
    parser.setInput(is, "UTF-8");
    int eventType = parser.getEventType();

    while (eventType != XmlPullParser.END_DOCUMENT) {
        switch (eventType) {
        case XmlPullParser.START_DOCUMENT:
            books = new ArrayList<Book>();
            break;
        case XmlPullParser.START_TAG:
            if (parser.getName().equals("book")) {
                book = new Book();
            } else if (parser.getName().equals("id")) {
                eventType = parser.next();
                book.setId(Integer.parseInt(parser.getText()));
            } else if (parser.getName().equals("name")) {
                eventType = parser.next();
                book.setName(parser.getText());
            } else if (parser.getName().equals("price")) {
                eventType = parser.next();
```

```
                book.setPrice(Float.parseFloat(parser.getText()));
            }
            break;
        case XmlPullParser.END_TAG:
            if (parser.getName().equals("book")) {
                books.add(book);
                book = null;
            }
            break;
        }
        eventType = parser.next();
    }

    return books;
}
```

文档刚被初始化时，事件为 START_DOCUMENT，可以通过 XmlPullParser.getEventType()来获取。然后调用 next()会产生 START_TAG，这个事件告诉应用程序一个标签已经开始了，调用 getName()会返回 book；若有 TEXT，则 next()会产生 TEXT 事件，调用 getText()会返回 TEXT。如果没有，则 next()会产生 END_TAG，说明一个标签已经处理完了。循环调用 next()直到最后处理完 TAG，会产生 END_DOCUMENT，说明整个文档已经处理完成了。除了 next()以外，nextToken()也可以使用，只不过它会返回更加详细的事件，例如 COMMENT、CDSECT、DOCDECL、ENTITY 等非常详细的信息。如果程序得到比较底层的信息，可以用 nextToken()来驱动并处理详细的事件。需要注意的一点是，TEXT 事件是有可能返回空白的(如换行符或空格等)。

nextTag()方法会忽略空白。如果可以确定下一个是 START_TAG 或 END_TAG，就可以调用 nextTag()直接跳过去。通常它有两个用处：当为 START_TAG 时，如果能确定这个 TAG 含有子 TAG，那么就可以调用 nextTag()产生子标签的 START_TAG 事件；当为 END_TAG 时，如果确定不是文档结尾，就可以调用 nextTag()产生下一个标签的 START_TAG。在这两种情况下，如果用 next()会有 TEXT 事件，但返回的是换行符或空白符。

nextText()方法只能在 START_TAG 时调用。当下一个元素是 TEXT 时，TEXT 的内容会返回；当下一个元素是 END_TAG 时，也就是说这个标签的内容为空，那么返回空字串；这个方法返回后，Parser 会停留在 END_TAG 上。

2.2 JSON 数据解析

JSON 为 Web 应用开发者提供了另一种数据交换格式，同 XML 或 HTML 片段相比，JSON 提供了更好的简单性和灵活性。

2.2.1 JSON 介绍

JSON(JavaScript Object Notation)是一种轻量级的数据交换格式,具有良好的可读性和便于快速编写的特性,同时也易于机器解析和生成,非常适合于服务器与客户端的交互。JSON 采用与编程语言无关的文本格式,业内主流技术为其提供了完整的解决方案(有点类似于正则表达式),从而可以在不同平台间进行数据交换。JSON 采用兼容性很高的文本格式[①],同时也具备类似于 C 语言体系的行为。

1. JSON 的数据结构

与 XML 一样,JSON 也是基于纯文本的数据结构(注意:JSON 并不是一个文档格式,没有 *.json 的文档,一般 JSON 格式的文档存在 txt 文件中)。

JSON 有以下两种数据结构。

(1) Map

Map 结构也称为对象,以键值对的形式给出,键和值之间用“:”隔开,两个 Map 之间用“,”隔开,一般表示形式如下:

```
{'key1':'value1','key2':'value2'}
```

图 2-1 演示了这种组织形式。

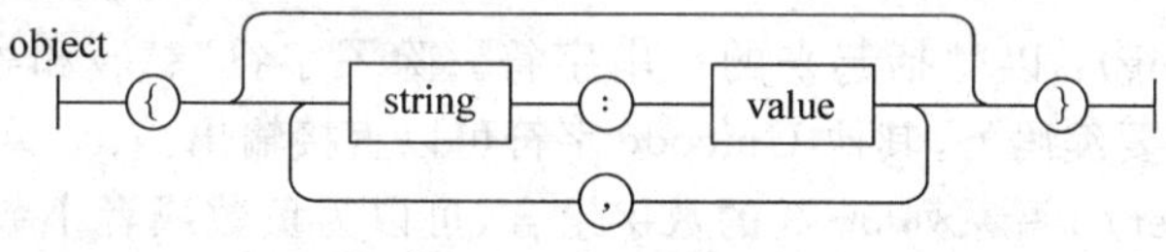

图 2-1 JSON Map 结构

(2) Array

Array 就是普通意义上的数组,一般形式如下:

```
['arr1','arr2','arr3'];
```

图 2-2 演示了这种组织形式。

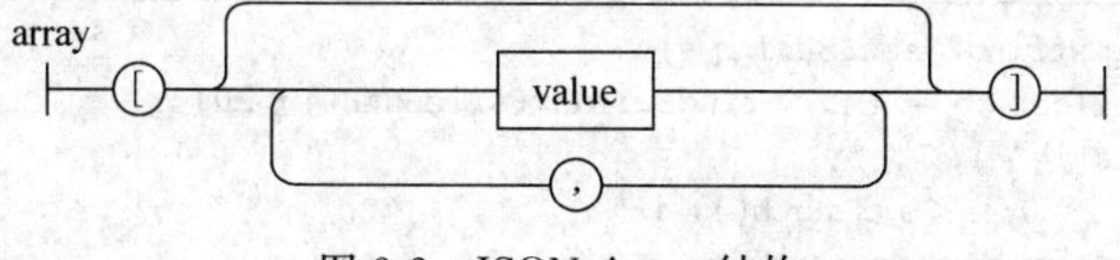

图 2-2 JSON Array 结构

说明:JSON 可以嵌套表示,例如 Array 中可以嵌套 Object。

2. JSON 的数据格式

由于 JSON 天生是为 JavaScript 准备的,因此,JSON 的数据格式非常简单,可以用 JSON 传输一个简单的 String、Number、Boolean,也可以传输一个数组,或者一个复杂的 Object 对象。

---

① http://www.json.org.cn/tools/JSONEditorOnline2.0/index.htm 提供了 JSON 在线编辑工具。

- 对象(Object)：一个对象以“{”开始，并以“}”结束。一个对象包含一系列非排序的键值对，每个键值对之间使用“,”分隔。此处的 Object 相当于 Java 中的 Map<String, Object>，而不是 Java 的 Class。例如，一个 Address 对象包含如下键值对：

```
{"city":"Beijing","street":" Chaoyang Road ","postcode":100025}
```

其中 Value 也可以是另一个 Object 或者数组。因此，复杂的 Object 可以嵌套表示，例如，一个 Person 对象包含 name 和 address 对象，可以表示如下：

```
{"name":"Michael","address": {"city":"Beijing","street":" Chaoyang Road ","postcode":100025}}
```

- 名称—值(Collection)：名称和值之间使用“:”隔开，一般的形式是：{name:value}。一个名称是一个字符串；一个值可以是一个字符串、一个数值、一个对象、一个布尔值、一个有序列表，或者一个 null 值，如{"firstName":"Brett"}。
- 数组(Array)：使用[ ]包含所有元素，每个元素用逗号分隔，元素可以是任意的 Value，例如，以下数组包含了一个 String、Number、Boolean 和一个 null：

```
[ "abc" , 12345 , false , null ]
```

- 字符串(String)：以""括起来的一串字符。除了字符"、\、/和一些控制符(\b、\f、\n、\r、\t)需要编码外，其他 Unicode 字符可以直接输出。
- 数值(Number)：一系列 0～9 的数字组合，可以为负数或者小数。还可以用 e 或者 E 表示为指数形式。
- 布尔值(Boolean)：表示为 true 或者 false。在很多语言中它被解释为数组。

注意：true、false 都没有双引号，否则将被视为一个字符串。

3. 生成 JSON

(1) 使用 String 生成 JSON

将 String 对象编码为 JSON 格式时，只需处理好特殊字符即可。另外，必须用双引号表示字符串，例如：

```
static String string2Json(String s) {
    StringBuilder sb = new StringBuilder(s.length() + 20);
    sb.append('\"');
    for (int i = 0; i < s.length(); i++) {
        char c = s.charAt(i);
        switch (c) {
        case '\"':
            sb.append("\\\"");
            break;
        case '\\':
            sb.append("\\\\");
            break;
        case '/':
            sb.append("\\/");
            break;
```

```
            case '\b':
                sb.append("\\b");
                break;
            case '\f':
                sb.append("\\f");
                break;
            case '\n':
                sb.append("\\n");
                break;
            case '\r':
                sb.append("\\r");
                break;
            case '\t':
                sb.append("\\t");
                break;
            default:
                sb.append(c);
            }
        }
        sb.append('\"');
        return sb.toString();
}
```

(2) 使用 Number 生成 JSON

利用 Java 的多态性，可以处理 Integer、Long、Float 等多种 Number 格式，例如：

```
static String number2Json(Number number) {
    return number.toString();
}
```

(3) 使用数组生成 JSON

通过循环将数组的每一个元素编码出来生成 JSON，例如：

```
static String array2Json(Object[] array) {
    if (array.length == 0)
        return "[]";
    StringBuilder sb = new StringBuilder(array.length << 4);
    sb.append('[');
    for (Object o : array) {
        sb.append(toJson(o));
        sb.append(',');
    }
    //将最后添加的 ',' 变为 ']':
    sb.setCharAt(sb.length() - 1, ']');
    return sb.toString();
}
```

其中的 toJson()方法代码如下：

```
public static String toJson(Object o) {
    if (o == null)
```

```
        return "null";
    if (o instanceof String)
        return string2Json((String)o);
    if (o instanceof Boolean)
        return boolean2Json((Boolean)o);
    if (o instanceof Number)
        return number2Json((Number)o);
    if (o instanceof Map)
        return map2Json((Map<String, Object>)o);
    if (o instanceof Object[])
        return array2Json((Object[])o);
    throw new RuntimeException("Unsupported type: "
            + o.getClass().getName());
}
```

(4) 使用 Map<String, Object>生成 JSON

与数组类似,通过遍历 Map 对象来生成 JSON,例如:

```
static String map2Json(Map<String, Object> map) {
    if (map.isEmpty())
        return "{}";
    StringBuilder sb = new StringBuilder(map.size() << 4);
    sb.append('{');
    Set<String> keys = map.keySet();
    for (String key : keys) {
        Object value = map.get(key);
        sb.append('\"');
        sb.append(key);
        sb.append('\"');
        sb.append(':');
        sb.append(toJson(value));
        sb.append(',');
    }
    //将最后的 ',' 变为 '}':
    sb.setCharAt(sb.length() - 1, '}');
    return sb.toString();
}
```

2.2.2 JSON 核心解析类

Android 的 JSON 解析部分都在 org.json 包下,主要有以下几个类。

1. JSONObject

JSONObject 是一个无序的键值对的集合,可以看作一个 JSON 对象,这是系统中有关 JSON 定义的基本单元。它的外在形式是一个用大括号包裹,并用冒号将名字和值分开的字符串。内部形式就是一个对象。

JSONObject 提供了一系列的 get、set 和 opt 方法来访问 JSONObject 实例。这些值的类型可以是 Boolean、JSONArray、JSONObject、Number、String 或者默认值 JSONObject.NULL 对象。

JSONObject 有两个不同的取值方法：

- getType 可以将要获取的键值转换为指定的类型，如果无法转换或没有值，则抛出 JSONException。
- optType 也是将要获取的键值转换为指定的类型，无法转换或没有值时，返回用户提供或者默认提供的值。

在使用 JSONObject 时，一般先创建一个 JSONObject 对象后，再使用 put(String, Object)等方法添加键值对，最后使用 toString()方法将 JSONObject 对象按照 JSON 的标准格式进行封装。

例如，下面的代码用于创建一个 JSONObject 对象：

```
private JSONObject createJSONObject() {
    JSONObject person = new JSONObject();
    try {
        JSONArray phone = new JSONArray();
        phone.put("12345678");
        phone.put("87654321");
        person.put("phone", phone);
        person.put("name", "liweiyong");
        person.put("age", 100);
        JSONObject address = new JSONObject();
        address.put("country", "china");
        address.put("province", "jiangsu");
        person.put("address", address);
        person.put("married", false);
    } catch (JSONException e) {
        e.printStackTrace();
    }
    return person;
}
```

创建的 JSON 对象形式如下：

```
{"phone":["12345678","87654321"],                          //数组
"married":false,                                          //布尔值
"address":{"province":"jiangsu","country":"china"},       //JSON 对象
"age":100,                                                //数值
"name":"liweiyong"                                        //字符串
}
```

2. JSONStringer

JSONStringer 是 JSON 文本构建类，用于帮助快速和便捷地创建 JSON 文本。其最大的优点在于可以减少由于格式的错误而导致程序异常，引用这个类可以自动严格按照 JSON 语法规则创建 JSON 文本。每个 JSONStringer 实体只能对应创建一个 JSON 文本。例如：

```
try {
    String myString = new JSONStringer().object()
        .key("id").value("20120226")
        .key("name").value("lxb")
        .endObject().toString();
} catch (JSONException ex) {
    throw new RuntimeException(ex);
}
```

结果是一组标准格式的 JSON 文本：{"id" ："20120226","name" ："lxb"}。为了按照 Object 标准给数值添加边界,.object()和.endObject()必须同时使用。同样,针对数组也有一组标准的方法.array()和.endArray()来生成边界。

3. JSONArray

JSONArray 代表一组有序的数值。表现形式是用方括号包裹,数值以“,”分隔(例如：[value1,value2,value3])。这个类的内部同样具有查询行为,通过 get()和 opt()两种方法都可以根据 index 索引返回指定的数值,put()方法用来添加或者替换数值。这个类和 JSONObject 支持相同的数据类型。

下面的代码是对 JSONArray 遍历的典型方式：

```
JSONArray array = JSONArray.fromObject(data);
for (Object object : array) {
    JSONObject o = JSONObject.fromObject(object);
    o.get("key")
}
```

4. JSONTokener

JSONTokener 是系统为 JSONObject 和 JSONArray 构造器的解析类,它可以从源信息中提取数值信息。例如：

```
private static String getJSONContent(){
    JSONTokener jsonTokener = new JSONTokener(JSONText);
    JSONObject studentJSONObject;
    String name = null;
    int id = 0;
    String phone = null;
    try {
        studentJSONObject = (JSONObject) jsonTokener.nextValue();
        name = studentJSONObject.getString("name");
        id = studentJSONObject.getInt("id");
        phone = studentJSONObject.getString("phone");
    } catch (JSONException e) {
        e.printStackTrace();
    }

    return name + " " + id + " " + phone;
}
```

5. SONException

JSONException 是 json. org 类抛出的异常信息。当语法错误或者过程异常的时候，会抛出 JSONException 异常。

以下情况下会产生 JSONException：

- 试图解析或构建一个格式错误的 JSON 文档。
- 使用 null 作为关键词。
- 使用时不提供给 JSON 数值类型，如 NaN 或无穷大的。
- 使用不存在的键进行查找。
- 类型不匹配的解析。

6. JSONString

JSONString 是一个接口，以便其他类可以通过实现该接口的 toString()方法来改变 JSONObject、JSONArray 等内部 toString()方法的功能，以实现它们自己的序列化。

7. JSONWriter

JSONWriter 位于 android. util 包下，是一个快速将 JSON 文本写入数据流的工具。每次只能输出一个字符串。流中既包括文字值(字符串、数字、布尔值和空值)，也包括作为对象、数组的开始和结束标志的分隔符。

例如，下面的 writeJsonStream()方法主要是利用 JsonWriter 把联系人的信息写入文件流中：

```
private void writeJsonStream() {
    ByteArrayOutputStream out = new ByteArrayOutputStream();
    JsonWriter writer = new JsonWriter(new OutputStreamWriter(out, "UTF - 8"));

    Cursor cur = context.getContentResolver().query(
            ContactsContract.Contacts.CONTENT_URI,
            null,
            null,
            null,
            ContactsContract.Contacts.DISPLAY_NAME
                    + " COLLATE LOCALIZED ASC");

    if (cur != null && cur.moveToFirst()) {
        int idColumn = cur.getColumnIndex(ContactsContract.Contacts._ID);

        int displayNameColumn = cur
                .getColumnIndex(ContactsContract.Contacts.DISPLAY_NAME);
        writer.setIndent(" ");
        writer.beginObject();
        writer.name(ContactStruct.CONTACTS);
        writer.beginArray();

        do {
            writer.beginObject();
            String contactId = cur.getString(idColumn);
            writer.name(ContactStruct._ID).value(contactId);
```

```
                String disPlayName = cur.getString(displayNameColumn);
                writer.name(ContactStruct.NAME).value(disPlayName);
                int phoneCount = cur.getInt(cur.getColumnIndex(
                        ContactsContract.Contacts.HAS_PHONE_NUMBER));
                Log.i("username", disPlayName);
                if (phoneCount > 0) {
                    Cursor phones = context.getContentResolver().query(
                            ContactsContract.CommonDataKinds.Phone.CONTENT_URI,
                            null,
                            ContactsContract.CommonDataKinds.Phone.CONTACT_ID
                                    + " = " + contactId, null, null);
                    if (phones != null && phones.moveToFirst()) {
                        writer.name(ContactStruct.PHONENUMBERS);
                        writer.beginArray();
                        do {
                            writer.beginObject();
                            String phoneNumber = phones.getString(phones
                                    .getColumnIndex(ContactsContract
                                            .CommonDataKinds.Phone.NUMBER));
                            String phoneType = phones.getString(phones
                                    .getColumnIndex(ContactsContract
                                            .CommonDataKinds.Phone.TYPE));
                            writer.name(ContactStruct.TYPE).value(phoneType);
                            writer.name(ContactStruct.PHONENUMBER).value(
                                    phoneNumber);
                            writer.endObject();
                            Log.i("phoneNumber", phoneNumber + "");
                            Log.i("phoneType", phoneType + "");
                        } while (phones.moveToNext());
                        writer.endArray();
                    }
                    closeCursor(phones);
                }
            } while (cur.moveToNext());
        }
}
```

8. JSONReader

JsonReader 位于 android.util 包下，主要用来读取 JSON 字符串的内容。例如，JSON 字符串如下：

```
private static final String JSONString =
        "[{\"name\":\"Michael\",\"age\":20},
        {\"name\":\"Mike\",\"age\":21}]";
```

在 Activity 中调用解析方法：

```
JSONUtils jsonUtils = new JSONUtils();
jSONUtils.parseJson(JSONString);
```

其中的 JSONUtils 代码如下：

```java
public class JSONUtils {
    public void parseJson(String jsonData) {
        try {
            JsonReader reader = new JsonReader(new StringReader(jsonData));
            reader.beginArray();
            while (reader.hasNext()) {
                reader.beginObject();
                while (reader.hasNext()) {
                    String tagName = reader.nextName();
                    if (tagName.equals("name")) {
                        System.out.println("name--->" + reader.nextString());
                    } else if (tagName.equals("age")) {
                        System.out.println("age--->" + reader.nextInt());
                    }
                }
                reader.endObject();
            }
            reader.endArray();
        } catch (Exception e) {
            e.printStackTrace();
        }
    }
}
```

在 LogCat 中观察运行结果，如图 2-3 所示。

```
System.out        name--->Michael
System.out        age--->20
System.out        name--->Mike
System.out        age--->21
```

图 2-3　JsonReader 解析结果

最后，演示一个通过向中国天气网发送获取天气信息的请求，然后解析获取的 JSON 数据并显示的示例。

首先创建一个如下的请求连接工具类 HttUtil，代码如下：

```java
public class HttUtil {

    public static HttpClient httpClient = new DefaultHttpClient();

    /**
     * @param url 请求地址
     * @return 服务器响应的字符串
     * @throws InterruptedException
     * @throws ExecutionException
     */
    public static String getRequest(final String url)
            throws InterruptedException, ExecutionException {
        FutureTask<String> task = new FutureTask<String>(
                new Callable<String>() {

```

```
                    @Override
                    public String call() throws Exception {
                        HttpGet get = new HttpGet(url);
                        HttpResponse httpResponse = httpClient.execute(get);
                        if (httpResponse.getStatusLine()
                                .getStatusCode() == 200) {
                            return EntityUtils.toString(httpResponse
                                    .getEntity());
                        }
                        return null;
                    }
                });

        new Thread(task).start();
        return task.get();
    }

}
```

然后实现如下简单的 Activity 实例：

```
public class JSONParsingActivity extends Activity {

    private TextView showResult;

    @Override
    protected void onCreate(Bundle savedInstanceState) {
        super.onCreate(savedInstanceState);
        setContentView(R.layout.activity_jsonparsing);

        showResult = (TextView) findViewById(R.id.result);

        WeatherAsyncTask task = new WeatherAsyncTask();
        task.execute("http://www.weather.com.cn/data/cityinfo/101010100.html");
    }

    class WeatherAsyncTask extends AsyncTask<String, Integer, String> {

        public WeatherAsyncTask() {
            super();
        }

        @Override
        protected String doInBackground(String... params) {
            String result = "";
            try {
                result = HttUtil.getRequest(params[0]);
            } catch (InterruptedException e) {
                e.printStackTrace();
```

```
29              } catch (ExecutionException e) {
30                  e.printStackTrace();
31              }
32
33              return result;
34          }
35
36          @Override
37          protected void onPostExecute(String result) {
38              try {
39                  //获取 JSONObject 对象
40                  JSONObject jsonobject = new JSONObject(result);
41                  JSONObject jsoncity = new JSONObject(
42                          jsonobject.getString("weatherinfo"));
43                  showResult.setText("城市:" + jsoncity.getString("city") + "\n"
44                          + "气温:最高" + jsoncity.getString("temp1")
45                          + "     最低" + jsoncity.getString("temp2") + "\n"
46                          + "今天天气:" + jsoncity.getString("weather"));
47              } catch (JSONException e) {
48                  e.printStackTrace();
49              }
50          }
51
52      }
53
54  }
```

代码解析如下：

12～13 行创建一个 16～52 行定义的 WeatherAsyncTask(继承自 AsyncTask)实例，并调用其 execute()方法执行异步任务。

22～34 行覆盖的 doInBackground()方法通过第 26 行来向中国天气网请求数据，并将数据赋给 result 字符串，该字符串的内容类似如下：

```
{"weatherinfo":{"city":"北京","cityid":"101010100","temp1":"15℃","temp2":"5℃","weather":"多云","img1":"d1.gif","img2":"n1.gif","ptime":"08:00"}}
```

36～50 行覆盖的 onPostExecute()方法对返回的结果 result 进行解析。通过上面的 result 结果可以看到，JSONObject 对象的一系列 get 方法里的参数 city、temp1、temp2、weather 等都是 key。

2.2.3 JSON 解析工具

1. Gson

Android 默认提供 JSONArray 和 JSONObject 来解析 JSON 格式的数据，但将 JSON 转换为实体对象时不是很方便。Gson[①] 是 Google 提供的一个轻量级的 JSON 转换类库，在

① gson.jar 下载地址：http://code.google.com/p/google-gson/downloads/list。

Java 平台可以方便地将一个 Java 对象转换成 JSON 格式,也可以将 JSON 格式的字符串转换成 Java 对象。

Gson 的解析非常简单,但是它的解析规则是必须有一个 Bean 文件,这个 Bean 文件的内容跟 JSON 数据类型是一一对应的。

假设 JSON 的数据格式如下:

```
{
    "id": 100,
    "body": "It is my post",
    "number": 0.13,
    "created_at": "2014 - 05 - 22 19:12:38"
}
```

那么只需要定义对应的一个 Bean 类,例如:

```
public class FooBean {
    public int id;
    public String body;
    public float number;
    public String created_at;
}
```

下面介绍 Gson 解析 JSON 的三个主要类。

(1) Gson: 使用 Gson 的主类,构造 Gson 类的实例后,可使用 toJson(Object)方法将 Bean 里面的内容转换为 JSON 内容,使用 fromJson(String, Class)方法将 JSON 对象封装出单个的 Bean 对象。

toJson()方法主要有以下几种形式。

- String toJson (JsonElement jsonElement): 用于将 JsonElement 对象(如 JsonObject、JsonArray 等)转换成 JSON 数据。
- String toJson(Object src): 用于将指定的 Object 对象序列化成相应的 JSON 数据。
- String toJson(Object src, Type typeOfSrc): 用于将指定的 Object 对象(可以包括泛型类型)序列化成相应的 JSON 数据。

fromJson()方法主要有以下几种形式:

- <T> T fromJson(JsonElement json, Class<T> classOfT)
- <T> T fromJson(JsonElement json, Type typeOfT)
- <T> T fromJson(JsonReader reader, Type typeOfT)
- <T> T fromJson(Reader reader, Class<T> classOfT)
- <T> T fromJson(Reader reader, Type typeOfT)
- <T> T fromJson(String json, Class<T> classOfT)
- <T> T fromJson(String json, Type typeOfT)

这些方法用于将不同形式的 JSON 数据解析成 Java 对象。例如:

```
public static final String JSON_DATA = "...";
FooBean foo = new Gson().fromJson(JSON_DATA, FooBean.class);
```

(2) GsonBuilder：用于创建 Gson 的实例。与使用 new Gson()不同的是，GsonBuilder 可进行与默认配置不同的相关设置。

例如，如果上例中的 created_at 定义为 Date 类型，有如下 Bean：

```
public class FooBean {
    public int id;
    public String body;
    public float number;
    public Date created_at;
}
```

解析时的方法如下：

```
public static final String JSON_DATA = "...";
GsonBuilder gsonBuilder = new GsonBuilder();
gsonBuilder.setDateFormat("yyyy-MM-dd HH:mm:ss");
Gson gson = gsonBuilder.create();
FooBean foo = gson.fromJson(JSON_DATA, FooBean.class);
```

(3) TypeToken：实现了获取泛型类型的功能，使用 Type = TypeToken(泛型){}.gettype()方法将会返回一个反射包下的 type 对象，这就是 fromJson()所需要的 type 类型。

例如，有 JSON 数组信息如下：

```
[{
    "id": 100,
    "body": "It is my post1",
    "number": 0.13,
    "created_at": "2014-05-20 19:12:38"
},
{
    "id": 101,
    "body": "It is my post2",
    "number": 0.14,
    "created_at": "2014-05-22 19:12:38"
}]
```

下面的代码实现了将其解析成数组：

```
public static final String JSON_DATA = "...";
FooBean[] foos = new Gson().fromJson(JSON_DATA, FooBean[].class);
```

下面的代码实现了将其解析成 List：

```
public static final String JSON_DATA = "...";
Type listType = new TypeToken<ArrayList<FooBean>>(){}.getType();
ArrayList<FooBean> foos = new Gson().fromJson(JSON_DATA, listType);
```

下面介绍使用 Gson 解析 2.2.2 小节天气预报的示例。

根据天气的 JSON 数据，首先建立 WeatherInfo 实体 Bean，部分代码如下：

```
public class WeatherInfo {
    private String city;
    private String cityid;
    private String temp1;
    private String temp2;
    private String weather;
    private String ptime;
    //省略 get 和 set 方法

    @Override
    public String toString() {
        return "WeatherInfo [city=" + city + ", cityid=" + cityid + ", temp1="
                + temp1 + ", temp2=" + temp2 + ", weather=" + weather
                + ", ptime=" + ptime + "]";
    }
}
```

建立 Mode 类 Weather，代码如下：

```
public class Weather {
    private WeatherInfo weatherinfo;

    public WeatherInfo getWeatherinfo() {
        return weatherinfo;
    }

    public void setWeatherInfo(WeatherInfo weatherinfo) {
        this.weatherinfo = weatherinfo;
    }
}
```

建立 View 类 Activity，代码如下：

```
public class GSONParsingActivity extends Activity {

    private TextView showResult;

    @Override
    protected void onCreate(Bundle savedInstanceState) {
        super.onCreate(savedInstanceState);
        setContentView(R.layout.activity_gsonparsing);
```

```

        showResult = (TextView) findViewById(R.id.result);

        WeatherAsyncTask task = new WeatherAsyncTask();
        task.execute("http://www.weather.com.cn/data/cityinfo/101010100.html");

    }

    private void handleWeatherResponse(String response) {
        try {
            Gson gson = new Gson();
            Weather weather = gson.fromJson(response, Weather.class);
            WeatherInfo info = weather.getWeatherinfo();
            showWeatherInfo(info);
        } catch (Exception e) {
        }
    }

    private void showWeatherInfo(WeatherInfo info) {

        showResult.setText("城市:" + info.getCity() + "\n"
                + "气温:最高" + info.getTemp1() + "最低" + info.getTemp2() + "\n"
                + "今天天气:" + info.getWeather());

    }

    class WeatherAsyncTask extends AsyncTask<String, Integer, String> {

        public WeatherAsyncTask() {
            super();
        }

        @Override
        protected String doInBackground(String... params) {
            String result = "";
            try {
                result = HttUtil.getRequest(params[0]);
            } catch (InterruptedException e) {
                e.printStackTrace();
            } catch (ExecutionException e) {
                e.printStackTrace();
            }

            return result;
        }

        @Override
        protected void onPostExecute(String result) {
            handleWeatherResponse(result);
```

```
        }

    }

}
```

2. Json-lib

Json-lib[①] 是一个 Java 类库,提供了如下功能:

- 将 Java beans、maps、collections、arrays 和 XML 转换成 JSON 格式的数据。
- 将 JSON 格式数据转换成 Java beans 对象。

使用 Json-lib 需要的 jar 包包括:

- jakarta commons-lang 2.5
- jakarta commons-beanutils 1.8.0
- jakarta commons-collections 3.2.1
- jakarta commons-logging 1.1.1
- ezmorph 1.0.6

下面是使用 Json-lib 实现的一个工具类的核心代码:

```
public class JsonUtil {

    /**
     * 设置日期转换格式
     */
    static {
        //注册器
        MorpherRegistry mr = JSONUtils.getMorpherRegistry();

        //可转换的日期格式,即 Json 串中可以出现以下格式的日期与时间
        DateMorpher dm = new DateMorpher(new String[] { Util.YYYY_MM_DD,
                Util.YYYY_MM_DD_HH_MM_ss, Util.HH_MM_ss, Util.YYYYMMDD,
                Util.YYYYMMDDHHMMSS, Util.HHMMss });
        mr.registerMorpher(dm);
    }

    /**
     * 从 json 串转换成实体对象
     * @param jsonObjStr e.g. {'name':'get','dateAttr':'2009-11-12'}
     * @param clazz Person.class
     * @return
     */
    public static Object getDtoFromJsonObjStr(String jsonObjStr, Class clazz) {
        return JSONObject.toBean(JSONObject.fromObject(jsonObjStr), clazz);
    }
```

① Json-lib 官网:http://json-lib.sourceforge.net/。

```
    /**
     * 从 json 串转换成实体对象,并且实体集合属性中存有另外的实体 Bean
     * @param jsonObjStr e.g. {'data':[{'name':'get'},{'name':'set'}]}
     * @param clazz e.g. MyBean.class
     * @param classMap e.g. classMap.put("data", Person.class)
     * @return Object
     */
    public static Object getDtoFromJsonObjStr(String jsonObjStr,
            Class clazz, Map classMap) {
        return JSONObject.toBean(JSONObject.fromObject(jsonObjStr),
                clazz, classMap);
    }

    /**
     * 把一个 json 数组串转换成普通数组
     * @param jsonArrStr e.g. ['get',1,true,null]
     * @return Object[]
     */
    public static Object[] getArrFromJsonArrStr(String jsonArrStr) {
        return JSONArray.fromObject(jsonArrStr).toArray();
    }

    /**
     * 把一个 json 数组串转换成实体数组
     * @param jsonArrStr e.g. [{'name':'get'},{'name':'set'}]
     * @param clazz e.g. Person.class
     * @return Object[]
     */
    public static Object[] getDtoArrFromJsonArrStr(String jsonArrStr,
            Class clazz) {
        JSONArray jsonArr = JSONArray.fromObject(jsonArrStr);
        Object[] objArr = new Object[jsonArr.size()];
        for (int i = 0; i < jsonArr.size(); i++) {
            objArr[i] = JSONObject.toBean(jsonArr.getJSONObject(i), clazz);
        }
        return objArr;
    }

    /**
     * 把一个 json 数组串转换成实体数组,且数组元素的属性中含有另外的实例 Bean
     * @param jsonArrStr
     * e.g. [{'data':[{'name':'get'}]},{'data':[{'name':'set'}]}]
     * @param clazz e.g. MyBean.class
     * @param classMap e.g. classMap.put("data", Person.class)
     * @return Object[]
     */
    public static Object[] getDtoArrFromJsonArrStr(String jsonArrStr,
            Class clazz, Map classMap) {
```

```
        JSONArray array = JSONArray.fromObject(jsonArrStr);
        Object[] obj = new Object[array.size()];
        for (int i = 0; i < array.size(); i++) {
            JSONObject jsonObject = array.getJSONObject(i);
            obj[i] = JSONObject.toBean(jsonObject, clazz, classMap);
        }
        return obj;
    }

    /**
     * 把一个 json 数组串转换成存放普通类型元素的集合
     * @param jsonArrStr e.g. ['get',1,true,null]
     * @return List
     */
    public static List getListFromJsonArrStr(String jsonArrStr) {
        JSONArray jsonArr = JSONArray.fromObject(jsonArrStr);
        List list = new ArrayList();
        for (int i = 0; i < jsonArr.size(); i++) {
            list.add(jsonArr.get(i));
        }
        return list;
    }

    /**
     * 把一个 json 数组串转换成集合,且集合里存放的是实例 Bean
     * @param jsonArrStr e.g. [{'name':'get'},{'name':'set'}]
     * @param clazz
     * @return List
     */
    public static List getListFromJsonArrStr(String jsonArrStr,
            Class clazz) {
        JSONArray jsonArr = JSONArray.fromObject(jsonArrStr);
        List list = new ArrayList();
        for (int i = 0; i < jsonArr.size(); i++) {
            list.add(JSONObject.toBean(jsonArr.getJSONObject(i), clazz));
        }
        return list;
    }

    /**
     * 把一个 json 数组串转换成集合,且集合里对象的属性中含有另外的实例 Bean
     * @param jsonArrStr
     * e.g. [{'data':[{'name':'get'}]},{'data':[{'name':'set'}]}]
     * @param clazz e.g. MyBean.class
     * @param classMap e.g. classMap.put("data", Person.class)
     * @return List
     */
    public static List getListFromJsonArrStr(String jsonArrStr,
            Class clazz, Map classMap) {
```

```
        JSONArray jsonArr = JSONArray.fromObject(jsonArrStr);
        List list = new ArrayList();
        for (int i = 0; i < jsonArr.size(); i++) {
            list.add(JSONObject.toBean(jsonArr.getJSONObject(i),
                    clazz, classMap));
        }
        return list;
    }

    /**
    * 把 json 对象串转换成 map 对象
    * @param jsonObjStr e.g. {'name':'get','int':1,'double',1.1,'null':null}
    * @return Map
    */
    public static Map getMapFromJsonObjStr(String jsonObjStr) {
        JSONObject jsonObject = JSONObject.fromObject(jsonObjStr);

        Map map = new HashMap();
        for (Iterator iter = jsonObject.keys(); iter.hasNext();) {
            String key = (String) iter.next();
            map.put(key, jsonObject.get(key));
        }
        return map;
    }

    /**
    * 把 json 对象串转换成 map 对象,且 map 对象里存放的是其他的实体 Bean
    * @param jsonObjStr e.g. {'data1':{'name':'get'},'data2':{'name':'set'}}
    * @param clazz e.g. Person.class
    * @return Map
    */
    public static Map getMapFromJsonObjStr(String jsonObjStr,
            Class clazz) {
        JSONObject jsonObject = JSONObject.fromObject(jsonObjStr);

        Map map = new HashMap();
        for (Iterator iter = jsonObject.keys(); iter.hasNext();) {
            String key = (String) iter.next();
            map.put(key, JSONObject.toBean(jsonObject.getJSONObject(key),
                    clazz));
        }
        return map;
    }

    /**
     * 把 json 对象串转换成 map 对象,且 map 对象里存放的是其他的实体 Bean,还含有另外的实体 Bean
     * @param jsonObjStr e.g. {'mybean':{'data':[{'name':'get'}]}}
     * @param clazz e.g. MyBean.class
```

```
     * @param classMap     e.g. classMap.put("data", Person.class)
     * @return Map
     */
    public static Map getMapFromJsonObjStr(String jsonObjStr,
                Class clazz, Map classMap) {
        JSONObject jsonObject = JSONObject.fromObject(jsonObjStr);

        Map map = new HashMap();
        for (Iterator iter = jsonObject.keys(); iter.hasNext();) {
            String key = (String) iter.next();
            map.put(key, JSONObject
                    .toBean(jsonObject.getJSONObject(key), clazz, classMap));
        }
        return map;
    }

    /**
     * 把实体 Bean、Map 对象、数组、列表集合转换成 json 串
     * @param obj
     * @return
     * @throws Exception String
     */
    public static String getJsonStr(Object obj) {
        String jsonStr = null;
        //配置 json
        JsonConfig jsonCfg = new JsonConfig();

        //注册日期处理器
        jsonCfg.registerJsonValueProcessor(java.util.Date.class,
                new JsonDateValueProcessor(Util.YYYY_MM_DD_HH_MM_ss));
        if (obj == null) {
            return "{}";
        }

        if (obj instanceof Collection || obj instanceof Object[]) {
            jsonStr = JSONArray.fromObject(obj, jsonCfg).toString();
        } else {
            jsonStr = JSONObject.fromObject(obj, jsonCfg).toString();
        }

        return jsonStr;
    }

    /**
     * 把 json 串、数组、集合(collection map)、实体 Bean 转换成 XML①
```

① XMLSerializer API：http://json-lib.sourceforge.net/apidocs/net/sf/json/xml/XMLSerializer.html。具体实例请参考：http://json-lib.sourceforge.net/xref-test/net/sf/json/xml/TestXMLSerializer_writes.html；http://json-lib.sourceforge.net/xref-test/net/sf/json/xml/TestXMLSerializer_writes.html。

```
     * @param obj
     * @return
     * @throws Exception String
     */
    public static String getXMLFromObj(Object obj) {
        XMLSerializer xmlSerial = new XMLSerializer();

        //配置 json
        JsonConfig jsonCfg = new JsonConfig();

        //注册日期处理器
        jsonCfg.registerJsonValueProcessor(java.util.Date.class,
                new JsonDateValueProcessor(Util.YYYY_MM_DD_HH_MM_ss));

        if ((String.class.isInstance(obj) &&
                String.valueOf(obj).startsWith("["))
                || obj.getClass().isArray()
                || Collection.class.isInstance(obj)) {
            JSONArray jsonArr = JSONArray.fromObject(obj, jsonCfg);
            return xmlSerial.write(jsonArr);
        } else {
            JSONObject jsonObj = JSONObject.fromObject(obj, jsonCfg);
            return xmlSerial.write(jsonObj);
        }
    }

    /**
     * 从 XML 中转换成 json 串
     * @param xml
     * @return String
     */
    public static String getJsonStrFromXML(String xml) {
        XMLSerializer xmlSerial = new XMLSerializer();
        return String.valueOf(xmlSerial.read(xml));
    }

}
```

除了前面介绍的 Gson、Json-lib 等解析工具外，FastJson[1]、Jackson[2] 也是常见的 JSON 解析工具。

2.3　习　　题

1. 从 asserts 中读取 XML 文件并解析显示在用户界面上。
2. 编程实现从中国天气预报网获取 json 天气信息，然后解析并显示在用户界面上。

---

① Fastjson 官网：http://code.alibabatech.com/wiki/display/FastJSON/Home-zh。
② Jackson 官网：http://jackson.codehaus.org。

第 3 章　网络连接与管理

移动互联网应用需要根据应用自身的特点和服务方式选择不同的联网策略，Android 平台为管理网络的连接类型和网络状态提供了健全的编程接口。

3.1　ConnectivityManager 与网络管理

Android SDK 为标准网络、bluetooth、NFC、Wi-Fi 等的连接与管理提供了丰富的接口。

3.1.1　ConnectivityManager 的功能

ConnectivityManager 是 android.net 包提供的用于管理与网络连接相关的操作类，包括查询网络连接状态、当网络状态发生改变时通知应用等。

ConnectivityManager 的主要任务包括：

- 监听手机网络的状态(包括 GPRS、Wi-Fi、UMTS 等)。
- 网络状态发生改变时发送广播。
- 当一个网络连接失败时进行故障切换(尝试连接到别的网络)。
- 为应用程序查询网络状态提供 API 接口。

初始化一个 ConnectivityManager 的方法如下：

```
ConnectivityManager connectivity = (ConnectivityManager) context
        .getSystemService(Context.CONNECTIVITY_SERVICE);
```

ConnectivityManager 提供的主要方法包括以下几点。

- NetworkInfogetActiveNetworkInfo()：获取当前连接可用的网络信息。
- NetworkInfo[]getAllNetworkInfo()：获取设备支持的所有网络类型的连接状态信息。
- NetworkInfogetNetworkInfo(int networkType)：获取特定网络类型的连接状态信息。
- intgetNetworkPreference()：检索当前的首选网络类型。
- boolean isActiveNetworkMetered()：确定当前网络是否计算流量。
- static booleanisNetworkTypeValid(int networkType)：判断给定的整数是否表示一种有效的网络类型。

注意：要执行和 ConnectivityManager 相关的网络操作，需要在应用程序的 AndroidManifest.xml 文件中包含以下权限。

```
<uses-permission android:name="android.permission.INTERNET" />
<uses-permission android:name="android.permission.ACCESS_NETWORK_STATE" />
```

如果需要更改网络状态,还需要包含如下权限:

```
<uses-permission android:name="android.permission.CHANGE_NETWORK_STATE" />
```

3.1.2 网络连接判断

1. 是否有网络

当应用程序试图连接到网络时,应该先检查网络连接是否可用。通常有如下两种做法。

方法1:

```
public static boolean isNetworkAvailable(Context context) {
    ConnectivityManager connectivity = (ConnectivityManager) context
            .getSystemService(Context.CONNECTIVITY_SERVICE);

    if (connectivity == null) {
        Log.w(LOG_TAG, "couldn't get connectivity manager");
    } else {
        NetworkInfo[] info = connectivity.getAllNetworkInfo();
        if (info != null) {
            for (int i = 0; i < info.length; i++) {
                if (info[i].isAvailable()) {
                    Log.d(LOG_TAG, "network is available");
                    return true;
                }
            }
        }
    }

    Log.d(LOG_TAG, "network is not available");
    return false;
}
```

方法2:

```
public static boolean isNetworkAvailable(Context context) {
    ConnectivityManager connectivity = (ConnectivityManager) context
            .getSystemService(Context.CONNECTIVITY_SERVICE);

    if (connectivity == null) {
        Log.w(LOG_TAG, "couldn't get connectivity manager");
    } else {
        NetworkInfo activeNetwork = connectivity.getActiveNetworkInfo();
    }

    return activeNetwork.isConnectedOrConnecting();
}
```

方法 1 中，08 行通过调用 ConnectivityManager 的 getAllNetworkInfo()方法返回 NetworkInfo 数组。NetworkInfo 描述给定类型的网络接口状态。NetworkInfo 类包含了对 Wi-Fi 和 MOBILE 两种网络模式连接的详细描述。

NetworkInfo 提供了以下重要的方法。

- getState()：获取代表着连接成功与否等状态的 State 对象。
- getDetailedState()：获取详细的状态信息。
- getExtraInfo()：获取附加信息。
- getReason()：获取连接失败的原因。
- getType()：获取网络类型(一般为 Wi-Fi 或 MOBILE)。
- getTypeName()：获取网络类型的名称(一般取值 Wi-Fi 或 MOBILE)。
- isAvailable()：判断该网络是否可用。
- isConnected()：判断是否已经连接。
- isConnectedOrConnecting()：判断是否已经连接或正在连接。
- isFailover()：判断是否连接失败。
- isRoaming()：判断是否漫游。

在 10～15 行对 NetworkInfo 数组的遍历中，通过调用 isAvailable()方法判断当前设备有无可用的网络。

方法 2 中，02～03 行获取 ConnectivityManager 实例，然后通过 08 行的 getActiveNetworkInfo()方法返回当前连接可用的网络信息，并调用 isConnectedOrConnecting()方法判断是否已经连接或正在连接可用网络。

2. 网络是否已连接

在确定有可用网络的前提下，还需要检查网络是否连接，方法如下：

```
public static boolean checkNetState(Context context) {
    boolean netstate = false;
    ConnectivityManager connectivity = (ConnectivityManager) context
            .getSystemService(Context.CONNECTIVITY_SERVICE);
    if (connectivity != null) {
        NetworkInfo[] info = connectivity.getAllNetworkInfo();
        if (info != null) {
            for (int i = 0; i < info.length; i++) {
                if (info[i].getState() == NetworkInfo.State.CONNECTED) {
                    netstate = true;
                    break;
                }
            }
        }
    }

    return netstate;
}
```

NetworkInfo 有两个枚举类型的成员变量 NetworkInfo. DetailedState 和 NetworkInfo . State,用于查看当前网络的状态。

3.1.3 网络接入类型

1. 网络接入类型

Android 平台手机主要有四种网络连接类型：无网络(这种状态可能是因为手机停机，网络没有开启，信号不好等原因)、使用 Wi-Fi 上网、CMWAP(中国移动代理)和 CMNET 上网。

下面的方法用于判断网络连接的类型：

```
public static int getNetworkState(Context context) {
    ConnectivityManager connManager = (ConnectivityManager) context
            .getSystemService(Context.CONNECTIVITY_SERVICE);

    //Wi-Fi
    State state = connManager.getNetworkInfo(ConnectivityManager.TYPE_WIFI)
            .getState();
    if (state == State.CONNECTED || state == State.CONNECTING) {
        return NETWORN_WIFI;
    }

    //3G
    state = connManager.getNetworkInfo(ConnectivityManager.TYPE_MOBILE)
            .getState();
    if (state == State.CONNECTED || state == State.CONNECTING) {
        return NETWORN_MOBILE;
    }

    return NETWORN_NONE;
}
```

ConnectivityManager 中的网络接入类型介绍如下。

- ConnectivityManager. TYPE_MOBILE：移动数据连接。当连接活跃时，所有数据流量将使用这个默认网络类型的接口(它有一个默认路由)，如 3G、GPRS 等。
- ConnectivityManager. TYPE_WIFI：Wi-Fi 网络。
- ConnectivityManager. TYPE_MOBILE_SUPL：一种基于标准、允许移动电话用户与定位服务器通信的协议。SUPL 的优势在于独立于运营商网络结构和性价比等。此外，与其他类型比较，基于 SUPL 的平台对现有运营商的网络影响较小。
- ConnectivityManager. TYPE_MOBILE_MMS：彩信网络。
- ConnectivityManager. TYPE_WiMAX：全球微波互联接入。这是一项新兴的宽带无线接入技术，能提供面向互联网的高速连接，数据传输距离最远可达 50km。WiMAX 还具有 QoS 保障、传输速率高、业务丰富多样等优点。WiMAX 的技术起点较高，采用了代表未来通信技术发展方向的 OFDM/OFDMA、AAS、MIMO 等先进技术。随着技术标准的发展，WiMAX 逐步实现宽带业务的移动化，而 3G 则实现移动业务的宽带化,两种网络的融合程度会越来越高。

- ConnectivityManager. TYPE_MOBILE_DUN: 提供了通过 Bluetooth 无线技术接入 Internet 和其他拨号服务的标准。最常见的情况是使用手机或笔记本电脑时以无线方式拨号接入 Internet。

2. CMWAP 和 CMNET

有的中国移动业务需要通过 CMWAP 接入点才能够连接网络,在做这类应用的时候,不可避免地需要判断当前 APN(APN 指一种网络接入技术,是通过手机上网时必须配置的一个参数,它决定了手机通过哪种接入方式来访问网络)并切换 APN,这样才能成功连接到网络,从而再进一步连接到服务器。

在移动网络中,目前国内主要有两种连网类型: CMWAP 和 CMNET。判断方法如下:

```
public static String getAPNType(Context context) {
    ConnectivityManager connMgr = (ConnectivityManager) context
            .getSystemService(Context.CONNECTIVITY_SERVICE);
    NetworkInfo networkInfo = connMgr.getActiveNetworkInfo();

    Log.d(LOG_TAG,
            "networkInfo.getExtraInfo() is " + networkInfo.getExtraInfo());
    if (networkInfo.getExtraInfo().toLowerCase().equals("cmnet")) {
        return "cmnet";
    } else {
        return "cmwap";
    }

    }
```

获取 APN 之后,需要判断是否是 CMWAP,如果不是则需要更改当前 APN 为 CMWAP,方法如下:

```
InetSocketAddress address;
address = new InetSocketAddress("10.0.0.172", 80);
java.net.Proxy proxy = new java.net.Proxy(
        java.net.Proxy.Type.HTTP, address);
HttpURLConnection conn = (HttpURLConnection)
        new URL(httpURL).openConnection(proxy);
```

3. 2G、3G 和 4G 的判断

在开发 Android 应用时,为了给用户节省流量,提高用户的体验度,还需要根据用户当前网络情况来做一些调整。

下面的方法用于判断网络的模式:

```
public static enum NetState {
    NET_NO, NET_2G, NET_3G, NET_4G, NET_WIFI, NET_UNKNOWN
};

```

```
05  public static NetState getNetTypeForWebView(Context context) {
06      NetState stateCode = NetState.NET_NO;
07      ConnectivityManager cm = (ConnectivityManager) context
08              .getSystemService(Context.CONNECTIVITY_SERVICE);
09      NetworkInfo ni = cm.getActiveNetworkInfo();
10
11      if (ni != null && ni.isConnectedOrConnecting()) {
12          switch (ni.getType()) {
13          case ConnectivityManager.TYPE_WIFI:
14              stateCode = NetState.NET_WIFI;
15              break;
16          case ConnectivityManager.TYPE_MOBILE:
17              switch (ni.getSubtype()) {
18              case TelephonyManager.NETWORK_TYPE_GPRS:     //联通 2G
19              case TelephonyManager.NETWORK_TYPE_CDMA:     //电信 2G
20              case TelephonyManager.NETWORK_TYPE_EDGE:     //移动 2G
21              case TelephonyManager.NETWORK_TYPE_1xRTT:
22              case TelephonyManager.NETWORK_TYPE_IDEN:
23                  stateCode = NetState.NET_2G;
24                  break;
25              case TelephonyManager.NETWORK_TYPE_EVDO_A:   //电信 3G
26              case TelephonyManager.NETWORK_TYPE_UMTS:
27              case TelephonyManager.NETWORK_TYPE_EVDO_0:
28              case TelephonyManager.NETWORK_TYPE_HSDPA:
29              case TelephonyManager.NETWORK_TYPE_HSUPA:
30              case TelephonyManager.NETWORK_TYPE_HSPA:
31              case TelephonyManager.NETWORK_TYPE_EVDO_B:
32              case TelephonyManager.NETWORK_TYPE_EHRPD:
33              case TelephonyManager.NETWORK_TYPE_HSPAP:
34                  stateCode = NetState.NET_3G;
35                  break;
36              case TelephonyManager.NETWORK_TYPE_LTE:
37                  stateCode = NetState.NET_4G;
38                  break;
39              default:
40                  stateCode = NetState.NET_UNKNOWN;
41              }
42              break;
43          default:
44              stateCode = NetState.NET_UNKNOWN;
45          }
46
47      }
48
49      return stateCode;
50  }
```

17 行通过 NetworksInfo 对象的 getSubType()和 getSubTypeName()可以获取对应的网络类型与名字。

18 行的 TelephonyManager 类主要提供了一系列用于访问与手机通信相关的状态和信息的 getter 方法,包括手机 SIM 的状态和信息、电信网络的状态及手机用户的信息等。

下面列出了常见的 getter 方法。

- getCallState():返回电话的状态,包括 CALL_STATE_IDLE(无任何状态)、ALL_STATE_OFFHOOK(接起电话时)和 CALL_STATE_RINGING(电话进来时)。
- getDataState():获取数据连接状态,包括 DATA_CONNECTED(已连接)、DATA_CONNECTING(正在连接)、DATA_DISCONNECTED(断开)和 DATA_SUSPENDED(暂停)。
- getDeviceId():返回当前移动终端的唯一标识,GSM 网络返回 IMEI,CDMA 网络返回 MEID。
- getNetworkOperator():返回 MCC+MNC 代码。
- getNetworkOperatorName():返回移动网络运营商的名字(SPN)。
- getNetworkType():获取网络类型,包括 NETWORK_TYPE_CDMA(网络类型为 CDMA)、NETWORK_TYPE_GPRS(网络类型为 GPRS)等。
- getSimSerialNumber():返回 SIM 卡的序列号(IMEI)。

3.1.4 监控网络连接状态

在 Android 网络应用程序开发中,经常要判断网络连接是否可用,因此经常有必要监听网络状态的变化。Android 的网络状态监听可以用 BroadcastReceiver 来接收网络状态改变的广播,具体实现步骤如下:

① 定义一个 BroadcastReceiver,并重载其中的 onReceive()方法,在其中完成监测连网状态的功能,例如:

```
private class ConnectionChangeReceiver extends BroadcastReceiver{
    @Override
    public void onReceive(Context context, Intent intent) {
        ConnectivityManager connectivityManager = (ConnectivityManager)
                context.getSystemService( Context.CONNECTIVITY_SERVICE );
        NetworkInfo activeNetInfo =
                connectivityManager.getActiveNetworkInfo();
        NetworkInfo mobNetInfo = connectivityManager.getNetworkInfo(
                ConnectivityManager.TYPE_MOBILE );
        if ( activeNetInfo != null ) {
            Toast.makeText( context, "Active Network Type : "
                    + activeNetInfo.getTypeName(),
                    Toast.LENGTH_SHORT ).show();
        }
        if( mobNetInfo != null ) {
            Toast.makeText( context, "Mobile Network Type : "
                    + mobNetInfo.getTypeName(),
                    Toast.LENGTH_SHORT ).show();
        }
    }
}
```

② 只要网络的连接状态发生变化，ConnectivityManager 就会立刻发送广播 CONNECTIVITY_ACTION。因此，需要在适当的地方注册 BroadcastReceiver，例如，在 Activity 的 onCreate()生命周期方法中添加如下动态注册信息：

```
IntentFilter intentFilter = new IntentFilter();
intentFilter.addAction(ConnectivityManager.CONNECTIVITY_ACTION);
registerReceiver(connectionChangeReceiver, intentFilter);
```

③ 在适当的地方取消注册 BroadcastReceiver，例如，在 Activity 的 onDestroye()生命周期方法中添加如下动态取消注册信息：

```
if (connectionChangeReceiver!= null) {
    unregisterReceiver(connectionChangeReceiver);
}
```

说明：通常网络的改变会比较频繁，因此没有必要不间断地注册监听网络的改变。另外，通常用户会在有 Wi-Fi 的时候进行下载动作，若切换到移动网络，则通常会暂停当前的下载，当监听到恢复到 Wi-Fi 的情况时，又开始恢复下载。

3.2 Wi-Fi 网络连接与管理

android.net.wifi 包提供了对 Wi-Fi 网络操作和管理的类，包括 WifiManager、ScanResult、WifiConfiguration、WifiInfo，此外还有 WifiLock、MulticastLock 等。

注意：要执行和 Wi-Fi 相关的网络操作，需要在应用程序的 AndroidManifest.xml 文件中包含以下权限：

```
<uses-permission android:name="android.permission.ACCESS_WIFI_STATE" />
<uses-permission android:name="android.permission.CHANGE_WIFI_STATE" />
```

3.2.1 WifiManager

WifiManager 提供 Wi-Fi 管理的各种 API，主要包含 Wi-Fi 的扫描、建立连接和配置等。

获取 WifiManager 实例的方法如下：

```
WifiManager mWifiManager = (WifiManager) context
        .getSystemService(Context.WIFI_SERVICE);
```

WifiManager 提供了如下的重要方法。

- addNetwork(WifiConfiguration config)：添加一个 config 描述的 Wi-Fi 网络，默认情况下，这个 Wi-Fi 网络是 DISABLE 状态的。
- enableNetwork(int netId, Boolean disableOthers)：连接 netId 所指的 Wi-Fi 网络，

并使其他网络都被禁用。

- disableNetwork(int netId)：让一个网络连接失效。
- removeNetwork(int netId)：移除某一个网络及其连接配置(与 disableNetwork()方法不同,disableNetwork()只是单纯地断开连接,保存的 ssid 和密码并不清除)。
- calculateSignalLevel(int rssi, int numLevels)：计算信号的等级。
- compareSignalLevel(int rssiA, int rssiB)：将连接 A 和连接 B 做对比。
- createWifiLock(int lockType, String tag)：创建一个 Wi-Fi 锁,锁定当前的 Wi-Fi 连接。
- updateNetwork(WifiConfiguration config)：更新一个网络连接的信息。
- setWifiEnabled()：让一个连接有效。
- isWifiEnabled()：判断一个 Wi-Fi 连接是否有效。
- disconnect()：断开连接。
- reconnect()：如果连接准备好了,则连通网络。
- getConfiguredNetworks()：获取网络连接的状态。
- getConnectionInfo()：获取当前连接的信息。
- getDhcpInfo()：获取 DHCP 的信息。
- getScanResulats()：获取扫描测试的结果。
- getWifiState()：获取一个 Wi-Fi 接入点是否有效。
- pingSupplicant()：ping 一个连接,判断是否能连通。
- ressociate()：即便连接没有准备好,也要连通。
- saveConfiguration()：保留一个配置信息。
- startScan()：开始扫描。

下面列举几个常见的应用。

(1) 开启 Wi-Fi 网络

```
public boolean openWifi() {
    boolean bRet = true;

    if (!mWifiManager.isWifiEnabled()) {
        bRet = mWifiManager.setWifiEnabled(true);
    }

    return bRet;
}
```

(2) 关闭 Wi-Fi 网络

```
public void closeWifi() {
    if (mWifiManager.isWifiEnabled()) {
        mWifiManager.setWifiEnabled(false);
    }
}
```

（3）检查 Wi-Fi 的状态

```
public int checkState() {
    return mWifiManager.getWifiState();
}
```

调用 setWifiEnabled()后，系统进行 Wi-Fi 模块的开启需要一定时间，此时通过 WifiManager. getWifiState()获取的状态来判断是否完成。这些状态包括以下几种。

- WifiManager. WIFI_STATE_DISABLED：Wi-Fi 网卡不可用。
- WifiManager. WIFI_STATE_DISABLING：Wi-Fi 网卡正在关闭。
- WifiManager. WIFI_STATE_ENABLED：Wi-Fi 网卡可用。
- WifiManager. WIFI_STATE_ENABLING：Wi-Fi 网卡正在打开。
- WifiManager. WIFI_STATE_UNKNOWN：未知网卡状态。

（4）断开指定的网络

```
public void disConnectionWifi(int netId) {
    mWifiManager.disableNetwork(netId);
    mWifiManager.disconnect();
}
```

此外 WifiManaer 还提供了一个内部的子类 WifiManagerLock。WifiManagerLock 的作用是在普通状态下，如果 Wi-Fi 的状态处于闲置，那么网络的连通将会暂时中断。但是如果把当前的网络状态锁上，那么 Wi-Fi 连通将会保持在一定状态，当然解除锁定之后，就会恢复常态。

示例代码如下：

```
WifiLock mWifiLock;
    //锁定 WifiLock
    public void acquireWifiLock() {
    mWifiLock.acquire();
    }

    //解锁 WifiLock
    public void releaseWifiLock() {
        //判断是否锁定
        if (mWifiLock.isHeld()) {
            mWifiLock.acquire();
        }
    }

    //创建一个 WifiLock
    public void createWifiLock() {
    mWifiLock = mWifiManager.createWifiLock("test");
}
```

3.2.2 ScanResult

ScanResult 主要通过 Wi-Fi 硬件的扫描来获取一些周边的 Wi-Fi 热点信息,包含 SSID、Capabilities、frequency、level(信号强度)等。

- SSID:网络的名字。当搜索一个网络时,就是靠这个来区分每个不同的网络接入点。
- Capabilities:网络接入的性能。这里主要用来判断网络的加密方式(WPA、WPE)等。
- Frequency:频率。每一个频道交互的 MHz 数。
- Level:等级。主要用来判断网络连接的优先数。

扫描网络的一般方式如下:

```
public void startScan() {
    mWifiManager.startScan();
    //得到扫描结果
    List<ScanResult> mWifiList = mWifiManager.getScanResults();
}
```

3.2.3 WifiConfiguration

WifiConfiguration 主要描述 Wi-Fi 配置的所有信息。

下面的方法用于连接一个新的 Wi-Fi 网络:

```
public boolean connect(String SSID, String Password, WifiCipherType Type) {
    if (!this.openWifi()) {
        return false;
    }

    while (mWifiManager.getWifiState() ==
            WifiManager.WIFI_STATE_ENABLING) {
        try {
            Thread.currentThread();
            Thread.sleep(100);
        } catch (InterruptedException ie) {
        }
    }

    WifiConfiguration wifiConfig = createWifiInfo(SSID, Password, Type);

    if (wifiConfig == null) {
        return false;
    }

    WifiConfiguration tempConfig = isExsits(SSID, mWifiManager);

```

```
23        if (tempConfig != null) {
24            mWifiManager.removeNetwork(tempConfig.networkId);
25        }
26
27        int netID = mWifiManager.addNetwork(wifiConfig);
28        boolean bRet = mWifiManager.enableNetwork(netID, false);
29
30        return bRet;
31    }
```

因为开启 Wi-Fi 功能需要一段时间(一般需要 1～3 秒),所以通过 06～13 行的等待使得 Wi-Fi 的状态变成 WIFI_STATE_ENABLED 的时候才能继续执行下面的语句。

15 行的 createWifiInfo()方法主要用于配置一个 Wi-Fi 网络,代码如下:

```
public WifiConfiguration createWifiInfo(String SSID, String Password,
        WifiCipherType Type) {
    WifiConfiguration config = new WifiConfiguration();
    config.allowedAuthAlgorithms.clear();
    config.allowedGroupCiphers.clear();
    config.allowedKeyManagement.clear();
    config.allowedPairwiseCiphers.clear();
    config.allowedProtocols.clear();
    config.SSID = "\"" + SSID + "\"";

    WifiConfiguration tempConfig = isExsits(SSID, mWifiManager);
    if (tempConfig != null) {
        mWifiManager.removeNetwork(tempConfig.networkId);
    }

    if (Type == WifiCipherType.WIFICIPHER_NOPASS) {
        config.wepKeys[0] = "";
        config.allowedKeyManagement.set(WifiConfiguration.KeyMgmt.NONE);
        config.wepTxKeyIndex = 0;
    }
    if (Type == WifiCipherType.WIFICIPHER_WEP) {
        config.hiddenSSID = true;
        config.wepKeys[0] = "\"" + Password + "\"";
        config.allowedAuthAlgorithms
                .set(WifiConfiguration.AuthAlgorithm.SHARED);
        config.allowedGroupCiphers
                .set(WifiConfiguration.GroupCipher.CCMP);
        config.allowedGroupCiphers
                .set(WifiConfiguration.GroupCipher.TKIP);
        config.allowedGroupCiphers
                .set(WifiConfiguration.GroupCipher.WEP40);
        config.allowedGroupCiphers
                .set(WifiConfiguration.GroupCipher.WEP104);
        config.allowedKeyManagement
```

```
35                  .set(WifiConfiguration.KeyMgmt.NONE);
36          config.wepTxKeyIndex = 0;
37      }
38      if (Type == WifiCipherType.WIFICIPHER_WPA) {
39          config.preSharedKey = "\"" + Password + "\"";
40          config.hiddenSSID = true;
41          config.allowedAuthAlgorithms
42                  .set(WifiConfiguration.AuthAlgorithm.OPEN);
43          config.allowedGroupCiphers
44                  .set(WifiConfiguration.GroupCipher.TKIP);
45          config.allowedKeyManagement
46                  .set(WifiConfiguration.KeyMgmt.WPA_PSK);
47          config.allowedPairwiseCiphers
48                  .set(WifiConfiguration.PairwiseCipher.TKIP);
49          //config.allowedProtocols.set(WifiConfiguration.Protocol.WPA);
50          config.allowedGroupCiphers
51                  .set(WifiConfiguration.GroupCipher.CCMP);
52          config.allowedPairwiseCiphers
53                  .set(WifiConfiguration.PairwiseCipher.CCMP);
54          config.status = WifiConfiguration.Status.ENABLED;
55      }
56      return config;
57  }
```

WifiConfiguration 包含 6 个子类。

- WifiConfiguration.AuthAlgorthm：用来判断加密方法。
- WifiConfiguration.GroupCipher：获取使用 GroupCipher 的方法来进行加密。
- WifiConfiguration.KeyMgmt：获取使用 KeyMgmt 进行的加密。
- WifiConfiguration.PairwiseCipher：获取使用 WPA 方式进行的加密。
- WifiConfiguration.Protocol：获取使用哪一种协议进行加密。

WifiConfiguration.Status：获取当前网络的状态。

代码的 24～35 行和 41～53 行就是这 6 个子类的设置方法。

54 行的 WifiConfiguration.Status 提供了 3 个字段表示 Wi-Fi 连接的三种状态：CURRENT(值为 0，表示处于连接状态)、DISABLED(值为 1，表示网络不可用)和 ENABLED(值为 2，表示网络可用但没连接)。

connect()方法第 21 行的 isExsits()方法用于判断当前的 Wi-Fi 网络是否存在，代码如下：

```
private static WifiConfiguration isExsits(String SSID,
        WifiManager wifiManager) {
    List<WifiConfiguration> existingConfigs = wifiManager
            .getConfiguredNetworks();

    for (WifiConfiguration existingConfig : existingConfigs) {
        if (existingConfig.SSID.equals("\"" + SSID + "\"")) {
```

```
                return existingConfig;
            }
        }

        return null;
}
```

3.2.4 WifiInfo

描述已建立网络连接后的 Wi-Fi 信息，包含 IP、Mac 地址、连接速度等信息。

下面的代码是获取 WifiInfo 实例的方法：

```
WifiInfo mWifiInfo = mWifiManager.getConnectionInfo();
```

WifiInfo 类的主要方法如下。

- getBSSID()：获取接入点的 Mac 地址。
- getIpAddress()：获取本机的 IP 地址。
- getLinkSpeed()：获取连接速度(不是下载速度)，单位为 Mbps。
- getMacAddress()：获取 Mac 地址。
- getNetworkId()：获取网络 ID 号。每一个设定好了的网络都有一个独一无二的整数型 ID 号，用来识别网络。
- getRssi()：返回收到的信号强度，是个负数(负数越大信号越好)。
- getSSID()：获取无线信号提供者的名称(就是要连接的网络的名字)。

下面的代码演示了如何使用这些方法：

```
public String getMacAddress() {
    return (mWifiInfo == null) ? "NULL" : mWifiInfo.getMacAddress();
}

public String getBSSID() {
    return (mWifiInfo == null) ? "NULL" : mWifiInfo.getBSSID();
}

public String getSSID() {
    return (mWifiInfo == null) ? "NULL" : mWifiInfo.getSSID();
}

public int getIpAddress() {
    return (mWifiInfo == null) ? 0 : mWifiInfo.getIpAddress();
}

public int getNetWordId() {
    return (mWifiInfo == null) ? 0 : mWifiInfo.getNetworkId();
}

```

```
public String getWifiInfo() {
    return (mWifiInfo == null) ? "NULL" : mWifiInfo.toString();
}
```

3.2.5 Wi-Fi Direct

Wi-Fi Direct 的 API 允许 Android 4.0(API level 14)以上的应用程序不通过网络或热点直接与周围的设备进行连接。应用程序可以迅速地查找附近的设备并交换信息。并且与蓝牙相比,Wi-Fi Direct 的通信范围更大。

1. WifiP2pManager

android.net.wifi.p2p.WifiP2pManager 类提供的方法可用于操作当前设备中的 Wi-Fi 硬件,实现诸如探测、连接对等设备等功能。

初始化一个 WifiP2pManager 的方法如下:

```
WifiP2pManager mManager = (WifiP2pManager)
        getSystemService(Context.WIFI_P2P_SERVICE);
```

WifiP2pManager 提供的主要方法包括以下几种。

- initialize():通过 Wi-Fi 框架对应用进行注册。这个方法必须在任何其他 Wi-Fi 直连方法使用之前调用。
- connect():用指定的配置来启动设备间的对等连接。
- cancelConnect():取消任何进行中的对等设备间连接请求。
- requestConnectInfo():获取一个设备的连接信息。
- createGroup():用当前设备作为组管理员来创建一个对等组。
- removeGroup():移除当前的点对点连接组。
- requestGroupInfo():获取点对点连接组的信息。
- discoverPeers():初始化对等设备的发现。
- requestPeers():获取当前发现的对等设备列表。

WifiP2pManager 类的方法需要传入一个监听器,以便 Wi-Fi Direct 框架能够通知 Activity 该调用的状态。WifiP2pManager 类的监听器接口见表 3-1。

表 3-1 WifiP2pManager 类的监听器接口

监听器接口	关联的方法
WifiP2pManager.ActionListener	connect() cancelConnect() createGroup() removeGroup() discoverPeers()
WifiP2pManager.ChannelListener	initialize()
WifiP2pManager.ConnectionInfoListener	requestConnectInfo()
WifiP2pManager.GroupInfoListener	requestGroupInfo()
WifiP2pManager.PeerListListener	requestPeers()

2. 实现 Wi-Fi Direct 连接

实现一个 Wi-Fi Direct 连接的基本步骤如下。

(1) 初始化设置

在使用 Wi-Fi Direct 的 API 之前,必须确保应用可以访问设备的硬件并且设备要支持 Wi-Fi Direct 的通信协议。

① 为设备的 Wi-Fi 硬件获取权限并在 Android 的清单文件中声明应用正确使用的最低 SDK 版本:

```
<uses-sdk android:minSdkVersion="14" />
<uses-permission android:name="android.permission.ACCESS_WIFI_STATE" />
<uses-permission android:name="android.permission.CHANGE_WIFI_STATE" />
<uses-permission android:name="android.permission.CHANGE_NETWORK_STATE" />
<uses-permission android:name="android.permission.INTERNET" />
<uses-permission android:name="android.permission.ACCESS_NETWORK_STATE" />
```

② 检查设备是否支持 Wi-Fi Direct 技术。一种好的解决办法是当广播接收器接收到一个 WIFI_P2P_STATE_CHANGED_ACTION 意图时,向 Activity 通知 Wi-Fi Direct 的状态和相应的反应。

例如:

```
@Override
public void onReceive(Context context, Intent intent) {
    ...
    String action = intent.getAction();
    if (WifiP2pManager.WIFI_P2P_STATE_CHANGED_ACTION.equals(action)) {
        int state = intent.getIntExtra(WifiP2pManager.EXTRA_WIFI_STATE, -1);
        if (state == WifiP2pManager.WIFI_P2P_STATE_ENABLED) {
            //Wifi Direct is enabled
        } else {
            //Wi-Fi Direct is not enabled
        }
    }
    ...
}
```

(2) 创建一个广播接收器

使用 Wi-Fi Direct 时,通过监听广播意图来获知某个特定事件的发生。为此,创建一个新的 BroadcastReceiver 类,用来监听系统的 Wi-Fi P2P 状态的改变。

一般构造方法如下:

```
public WiFiDirectBroadcastReceiver(WifiP2pManager manager, Channel channel,
        WifiDirectActivity activity) {
    super();
    this.manager = manager;
    this.channel = channel;
```

```
    this.activity = activity;
}
```

构造方法一般最常用的就是以 WifiP2pManager、WifiP2pManager.Channel 作为参数，同时这个广播接收器对应的 Activity 也将被注册进来。这个广播接收器可以向 Activity 发送更新或者在需要的时候可以访问 Wi-Fi 硬件或通信通道。

在 onReceive()方法中,添加一个条件来处理各种 P2P 状态的变更。代码如下：

```
@Override
public void onReceive(Context context, Intent intent) {
    String action = intent.getAction();
    if (WifiP2pManager.WIFI_P2P_STATE_CHANGED_ACTION.equals(action)) {
        //确定 Wi-Fi Direct 模式是否已经启用,并提醒 Activity
        int state = intent.getIntExtra(WifiP2pManager.EXTRA_WIFI_STATE, -1);
        if (state == WifiP2pManager.WIFI_P2P_STATE_ENABLED) {
            activity.setIsWifiP2pEnabled(true);
        } else {
            activity.setIsWifiP2pEnabled(false);
        }
    } else if (WifiP2pManager.WIFI_P2P_PEERS_CHANGED_ACTION
            .equals(action)) {
        //对等点列表已经改变
    } else if (WifiP2pManager.WIFI_P2P_CONNECTION_CHANGED_ACTION
            .equals(action)) {
        //连接状态已经改变
    } else if (WifiP2pManager.WIFI_P2P_THIS_DEVICE_CHANGED_ACTION
            .equals(action)) {
        DeviceListFragment fragment = (DeviceListFragment)
            activity.getFragmentManager()
                .findFragmentById(R.id.frag_list);
        fragment.updateThisDevice((WifiP2pDevice)
            intent.getParcelableExtra(
                WifiP2pManager.EXTRA_WIFI_P2P_DEVICE));
    }
}
```

其中,监听动作包括以下几种。

- WIFI_P2P_STATE_CHANGED_ACTION：表明 Wi-Fi 对等网络(P2P)是否已经启用。
- WIFI_P2P_PEERS_CHANGED_ACTION：表明可用的对等点的列表发生了改变(调用 discoverPeers()方法时)。
- WIFI_P2P_CONNECTION_CHANGED_ACTION：表示 Wi-Fi 对等网络的连接状态发生了改变。
- WIFI_P2P_THIS_DEVICE_CHANGED_ACTION：表示该设备的配置信息发生了改变。

(3) 创建一个对等网络管理器

在应用程序的 Activity 中，初始化一个 IntentFilter，代码如下：

```
private final IntentFilter intentFilter = new IntentFilter();
intentFilter.addAction(WifiP2pManager.WIFI_P2P_STATE_CHANGED_ACTION);
intentFilter.addAction(WifiP2pManager.WIFI_P2P_PEERS_CHANGED_ACTION);
intentFilter.addAction(WifiP2pManager.WIFI_P2P_CONNECTION_CHANGED_ACTION);
intentFilter.addAction(WifiP2pManager.WIFI_P2P_THIS_DEVICE_CHANGED_ACTION);
```

然后在 onCreate()方法中初始化 WifiP2pManager 的一个实例，并调用它的 initialize()方法。这个方法返回一个 WifiP2pManager. Channel 对象。WifiP2pManager. Channel 用于把应用程序连接到 Wi-Fi Direct 框架中。代码如下：

```
WifiP2pManager mManager = (WifiP2pManager)
        getSystemService(Context.WIFI_P2P_SERVICE);
Channel mChannel = mManager.initialize(this, getMainLooper(), null);
BroadcastReceiver mReceiver = new WiFiDirectBroadcastReceiver(mManager,
        mChannel, this);
```

最后在 onResume()和 onPause()方法中注册和取消注册监听广播，代码如下：

```
@Override
public void onResume() {
    super.onResume();
    registerReceiver(receiver, intentFilter);
}

@Override
public void onPause() {
    super.onPause();
    unregisterReceiver(receiver);
}
```

(4) 发现对等设备

调用 discoverPeers()方法去发现可以使用并连接的对等设备。这个方法的调用是异步的。同时如果创建了一个 WifiP2pManager. ActionListener 监听器，可以通过 onSuccess()或者 onFailure()方法收到发现成功或失败的消息。onSuccess()方法只能通知发现的过程是否成功而不能提供任何关于发现设备的信息，例如：

```
mManager.discoverPeers(channel, new WifiP2pManager.ActionListener() {
    @Override
    public void onSuccess() {
        ...
    }

    @Override
```

```
        public void onFailure(int reasonCode) {
            ...
        }
    }
);
```

如果发现过程成功且检测到了对等设备,系统将会广播出一个 WIFI_P2P_PEERS_CHANGED_ACTION 意图,这样就可以利用广播监听器监听并获得发现设备的列表。当应用接收到 WIFI_P2P_PEERS_CHANGED_ACTION 意图时,可以调用 requestPeers()方法来获取发现设备的列表,代码如下:

```
private List peers = new ArrayList();
...
if (WifiP2pManager.WIFI_P2P_PEERS_CHANGED_ACTION.equals(action)) {

    //request available peers from the wifi p2p manager. This is an
    //asynchronous call and the calling activity is notified with a
    //callback on PeerListListener.onPeersAvailable()
    if (manager != null) {
        manager.requestPeers(channel, peerListListener);
    }
    }

    private PeerListListener peerListListener = new PeerListListener() {
    @Override
    public void onPeersAvailable(WifiP2pDeviceList peerList) {
        peers.clear();
        peers.addAll(peerList.getDeviceList());

        //如果 AdapterView 可以处理该数据,则把变更通知它
        ((WiFiPeerListAdapter) getListAdapter()).notifyDataSetChanged();
        if (peers.size() == 0) {
            Log.d(WiFiDirectActivity.TAG, "No devices found");
            return;
        }
    }
}
```

(5) 连接到设备

在获得发现设备列表之后,调用 connect()方法去连接指定设备。这个方法的调用需要一个包含待连接设备信息的 WifiP2pConfig 对象。可以通过 WifiP2pManager.ActionListener 接收到连接是否成功的通知。下面的代码展示了连接过程:

```
@Override
public void connect() {
    //使用在网络上找到的第一个设备
    WifiP2pDevice device = peers.get(0);
```

```

    WifiP2pConfig config = new WifiP2pConfig();
    config.deviceAddress = device.deviceAddress;
    config.wps.setup = WpsInfo.PBC;

    mManager.connect(mChannel, config, new ActionListener() {

        @Override
        public void onSuccess() {
            //WiFiDirectBroadcastReceiver 发出通知
        }

        @Override
        public void onFailure(int reason) {
            Toast.makeText(WiFiDirectActivity.this, "Connect failed. Retry."
                    Toast.LENGTH_SHORT).show();
        }
    });
}
```

在这个代码片段中实现的 WifiP2pManager.ActionListener 只会在初始化成功或失败时通知用户。要监听连接状态的变更，需要实现 WifiP2pManager.ConnectionInfoListener 接口。其回调方法 onConnectionInfoAvailable()将会在连接状态改变时发出通知。对于多个设备连接一个设备的情况（比如，多于 3 个玩家的游戏，或者聊天软件），其中一个设备将会被指定为“群主”。

```
@Override
public void onConnectionInfoAvailable(final WifiP2pInfo info) {

    //InetAddress 在 WifiP2pInfo 结构体中
    InetAddress groupOwnerAddress =
            info.groupOwnerAddress.getHostAddress());

    //组群协商后，就可以确定群主
    if (info.groupFormed && info.isGroupOwner) {
        //针对群主做某些任务
        //一种常用的做法是，创建一个服务器线程并接收连接请求
    } else if (info.groupFormed) {
        //其他设备都作为客户端。在这种情况下，会希望创建一个客户端线程来连接群主
    }
}
```

同时，修改 BroadcastReceiver 的 onReceive()方法，其中 WIFI_P2P_CONNECTION_CHANGED_ACTION 的部分代码如下：

```
if (WifiP2pManager.WIFI_P2P_CONNECTION_CHANGED_ACTION.equals(action)) {

```

```
    if (mManager == null) {
        return;
    }

    NetworkInfo networkInfo = (NetworkInfo) intent
            .getParcelableExtra(WifiP2pManager.EXTRA_NETWORK_INFO);

    if (networkInfo.isConnected()) {

        //连上了其他的设备,请求连接信息,以找到群主的 IP
        mManager.requestConnectionInfo(mChannel, connectionListener);
    }
}
```

其中的 requestConnectionInfo()是一个异步的调用,所以结果会传给作为参数的连接信息监听器。

(6) 数据传输

一旦连接建立,就可以通过 Socket 来进行数据的传输。基本的数据传输步骤如下:

① 创建一个 ServerSocket 对象。这个服务端套接字对象等待一个来自指定地址和端口的客户端连接且阻塞线程直到连接发生,所以把它建立在一个后台线程里。

② 创建一个客户端 Socket 对象。这个客户端套接字对象使用指定 IP 地址和端口去连接服务端设备。

③ 从客户端向服务端发送数据。当客户端成功连接服务端设备后,可以通过字节流从客户端向服务端发送数据。

④ 服务端等待客户端的连接(使用 accept()方法)。这个调用阻塞服务端线程直到客户端连接成功。当连接建立时,服务端可以接受来自客户端的数据。执行关于数据的任何动作,如保存数据或者展示给用户。

下面的例子,展示了怎样去创建服务端和客户端的连接和通信,并且使用一个客户端到服务端的服务来传输了一张图像。

```
public static class FileServerAsyncTask extends AsyncTask {

    private Context context;
    private TextView statusText;

    public FileServerAsyncTask(Context context, View statusText) {
        this.context = context;
        this.statusText = (TextView) statusText;
    }

    @Override
    protected String doInBackground(Void... params) {
        try {
```

```

            /**
             * Create a server socket and wait for client connections. This
             * call blocks until a connection is accepted from a client
             */
            ServerSocket serverSocket = new ServerSocket(8888);
            Socket client = serverSocket.accept();

            /**
             * If this code is reached, a client has connected and transferred
             * data Save the input stream from the client as a JPEG file
             */
            final File f = new File(Environment.getExternalStorageDirectory()
                    + "/"
                    + context.getPackageName()
                    + "/wifip2pshared-"
                    + System.currentTimeMillis()
                    + ".jpg");

            File dirs = new File(f.getParent());
            if (!dirs.exists())
                dirs.mkdirs();
            f.createNewFile();
            InputStream inputstream = client.getInputStream();
            copyFile(inputstream, new FileOutputStream(f));
            serverSocket.close();
            return f.getAbsolutePath();
        } catch (IOException e) {
            Log.e(WiFiDirectActivity.TAG, e.getMessage());
            return null;
        }
    }

    /**
     * Start activity that can handle the JPEG image
     */
    @Override
    protected void onPostExecute(String result) {
        if (result != null) {
            statusText.setText("File copied - " + result);
            Intent intent = new Intent();
            intent.setAction(android.content.Intent.ACTION_VIEW);
            intent.setDataAndType(Uri.parse("file://" + result), "image/*");
            context.startActivity(intent);
        }
    }
}
```

在客户端,使用客户端套接字连接服务端套接字并传输数据。核心代码如下:

```
Context context = this.getApplicationContext();
    String host;
    int port;
    int len;
    Socket socket = new Socket();
    byte buf[] = new byte[1024];
    ...
    try {
    /**
     * Create a client socket with the host,
     * port, and timeout information.
     */
    socket.bind(null);
    socket.connect((new InetSocketAddress(host, port)), 500);

    /**
     * Create a byte stream from a JPEG file and pipe it to the output stream
     * of the socket. This data will be retrieved by the server device.
     */
    OutputStream outputStream = socket.getOutputStream();
    ContentResolver cr = context.getContentResolver();
    InputStream inputStream = null;
    inputStream = cr.openInputStream(Uri.parse("path/to/picture.jpg"));
    while ((len = inputStream.read(buf)) != -1) {
        outputStream.write(buf, 0, len);
    }
    outputStream.close();
    inputStream.close();
    } catch (FileNotFoundException e) {
    //catch logic
    } catch (IOException e) {
    //catch logic
    }

    /**
     * Clean up any open sockets when done
     * transferring or if an exception occurred.
     */
    finally {
    if (socket != null) {
        if (socket.isConnected()) {
            try {
                socket.close();
            } catch (IOException e) {
                //catch logic
            }
        }
    }
}
```

3. Wi-Fi Direct 服务搜索

使用 Wi-Fi Direct 服务搜索，可以在不连接网络的情况下，直接找到附近的设备，也可以通知哪些服务在本地机器上运行。这些功能可以实现在没有可用的本地网络或热点时，进行应用程序之间的通信。

(1) 添加一个本地服务

如果要提供一个本地服务，需要注册服务搜索，然后框架会自动向对等点的服务搜索请求做出回应。创建一个本地服务的步骤如下：

① 创建一个 WifiP2pServiceInfo 对象。

② 将服务信息填入该对象。

③ 调用 addLocalService()来注册用于服务搜索的本地服务。

例如：

```
private void startRegistration() {
    //创建一个包含服务信息的字符串映射
    Map record = new HashMap();
    record.put("listenport", String.valueOf(SERVER_PORT));
    record.put("buddyname", "John Doe" + (int) (Math.random() * 1000));
    record.put("available", "visible");

    //服务信息。将一个实例名传给它,服务类型为_protocol._transportlayer。映射包含其他设备在连上它时希望获取的信息

    WifiP2pDnsSdServiceInfo serviceInfo = WifiP2pDnsSdServiceInfo
            .newInstance("_test", "_presence._tcp", record);

    //添加本地服务,用来发送服务信息,网络频道,以及将会用到的用来指示请求成功的监听器

    mManager.addLocalService(channel, serviceInfo, new ActionListener() {
        @Override
        public void onSuccess() {
            //命令成功
        }

        @Override
        public void onFailure(int arg0) {
            //命令失败。检查 P2P_UNSUPPORTED,ERROR 或者 BUSY
        }
    });
}
```

P2P_UNSUPPORTED 表示运行程序的设备不支持 Wi-Fi Direct；BUSY 表示系统太忙，无法处理请求；ERROR 表示内部错误引起的操作失败。

(2) 搜索附近的服务

Android 使用回调方法来通知应用程序有哪些可用的服务，因此首先要创建一个 WifiP2pManager.DnsSdTxtRecordListener 来监听进来的记录。这个记录也可以被其他设

备广播。当有某个设备加入时,将设备地址和其他需要的信息复制到当前方法以外的一个数据结构中,这样就可以稍后再对它进行操作。下面的例子假定记录包含一个 buddyname 域,填入的是用户身份。

```
final HashMap<String, String> buddies = new HashMap<String, String>();
    ...
    private void discoverService() {
    DnsSdTxtRecordListener txtListener = new DnsSdTxtRecordListener() {
        @Override
        /* 回调包含以下选项。
         * fullDomain: 完整的域名: 例如"printer._ipp._tcp.local"
         * record: 成对记录有关键字和值的数据的 TXT
         * device: 运行通知服务的设备
         */
        public void onDnsSdTxtRecordAvailable(
                String fullDomain, Map record, WifiP2pDevice device) {
                Log.d(TAG, "DnsSdTxtRecord available -" + record.toString());
                buddies.put(device.deviceAddress, record.get("buddyname"));
            }
        };
    ...
}
```

(3) 获取服务信息

要获得服务信息,需要创建一个 WifiP2pManager. DnsSdServiceResponseListener。它接收实际的描述和连接信息。前面的代码片段实现了一个 Map 对象,将设备地址和其伙伴名称进行配对。服务回应监听器使用这个对象来将 DNS 记录和对应的服务信息联系起来。当两个监听器都实现了,就用 setDnsSdResponseListeners()方法把它们添加进 WifiP2pManager 中。

例如:

```
private void discoverService() {
    ...

    DnsSdServiceResponseListener servListener = new
            DnsSdServiceResponseListener() {
        @Override
        public void onDnsSdServiceAvailable(String instanceName,
                String registrationType, WifiP2pDevice resourceType) {

                //假定有一个设备加入了,就使用 DnsTxtRecord 中的通俗易懂的信息更新设
                  备名称

                resourceType.deviceName = buddies
                        .containsKey(resourceType.deviceAddress) ? buddies
                        .get(resourceType.deviceAddress) :
                            resourceType.deviceName;
```

```

                //添加到为显示 Wi-Fi 设备而自定义的适配器中
                WiFiDirectServicesList fragment = (WiFiDirectServicesList)
                        getFragmentManager()
                        .findFragmentById(R.id.frag_peerlist);
                WiFiDevicesAdapter adapter = ((WiFiDevicesAdapter)
                        fragment.getListAdapter());

                adapter.add(resourceType);
                adapter.notifyDataSetChanged();
                Log.d(TAG, "onBonjourServiceAvailable " + instanceName);
        }
    };

    mManager.setDnsSdResponseListeners(channel, servListener, txtListener);
    ...
}
```

现在创建一个服务请求,并调用 addServiceRequest()。这个方法也使用一个监听器来汇报成功或失败。

```
serviceRequest = WifiP2pDnsSdServiceRequest.newInstance();
mManager.addServiceRequest(channel, serviceRequest, new ActionListener() {
    @Override
    public void onSuccess() {
        //成功
    }

    @Override
    public void onFailure(int code) {
        //命令失败
    }
});
```

最后,调用 discoverServices()方法,例如:

```
mManager.discoverServices(channel, new ActionListener() {

    @Override
    public void onSuccess() {
        //成功

    }

    @Override
    public void onFailure(int code) {
        //命令失败。检查 P2P_UNSUPPORTED、ERROR 或者 BUSY
        if (code == WifiP2pManager.P2P_UNSUPPORTED) {
```

```
            Log.d(TAG, "P2P isn't supported on this device.");
        else if(...)
            ...
    }
});
```

在异步调用中,包含一个 WifiP2pManager. ActionListener 参数,它提供了指明成功或失败的回调方法。如需诊断问题,将调试代码写在 onFailure()里面。

3.3 网络服务优化

本节给出一些关于网络连接和数据传输的优化建议及措施。

3.3.1 网络连接的优化

网络连接往往是耗电量比较大的,有效的网络连接对应用 App 的性能有重要的影响。下面是一些对网络连接的优化建议:

(1) 在需要网络连接的程序中,首先检查网络连接是否正常。如果没有网络连接,那么就不需要执行相应的程序。

(2) 应用程序没有必要不间断地注册监听网络的改变,可以在执行一个单元操作时监听网络的状态。例如,在执行一个 Wi-Fi 模式下的下载动作时,若 Wi-Fi 网络切换到移动网络,则通常会暂停当前下载,直到执行另一个任务时,监听到恢复到 Wi-Fi 时,再开始恢复下载。

(3) Wi-Fi 比蜂窝数据更省电,包括 2G(GPRS)、3G。尽量在 Wi-Fi 下传输数据(通过提前加载数据,可以减少下载数据所需要的无线连接数)。而在非 Wi-Fi 下,尽量减少网络访问。虽然 Wi-Fi 接入方式已经占到移动互联网用户的 50%,但是有些手机设置为待机时关闭 Wi-Fi 连接,即便有 Wi-Fi 信号,也只能切换到蜂窝数据。

(4) 通常来说,使用已经存在的网络连接会比重新初始化一个新的网络连接更有效率。重复使用连接还能使网络对于拥塞和相关的网络数据问题的响应更加智能。例如,应该将请求捆绑到一个 GET 当中,而不是创建多个连接同时下载数据,或多个连接连续地发送 GET 请求。

3.3.2 数据传输的优化

在 2009 年 Google IO 大会上,Jeffrey Sharkey 的演讲(Coding for Life — Battery Life, That Is)中就探讨了 Android 应用耗电的三个方面:大数据量的传输、不停地在网络间切换和解析大量的文本数据。

下面是一些对网络传输的优化建议:

(1) 使用效率高的数据格式和解析方法,推荐使用 JSON 和 Protobuf。

(2) 在进行大数据量下载时,尽量使用 GZIP 方式。

(3) 减少下载的最基本方法就是只下载所需的。在数据方面通过实现 REST APIs,就可以指定查询条件来限制返回的数据。

(4) 使用本地缓存文件,避免下载重复数据。通常缓存静态资源,包括按需下载的合理资源,并将这些资源单独存储,这样就可以定期清理缓存以保证缓存的大小。

(5) 每次数据传输的时候尽可能地保持当前的连接模式不改变,更不要频繁地切换模式。

(6) 使用预取模式下载大数据。预取数据是一种减少独立数据传输会话数量的有效方法。预取技术允许通过一次连接,最大限度地下载到给定时间内单次操作所需的所有数据。通过预先加载传输,可以减少下载数据所需的电量损耗,而且改善了延迟,降低了带宽占用,减少了下载次数。

下面的方法实现了根据不同的网络类型来设置不同的预取数据量:

```
ConnectivityManager cm = (ConnectivityManager)getSystemService(
        Context.CONNECTIVITY_SERVICE);

TelephonyManager tm = (TelephonyManager)getSystemService(
        Context.TELEPHONY_SERVICE);

NetworkInfo activeNetwork = cm.getActiveNetworkInfo();

int PrefetchCacheSize = DEFAULT_PREFETCH_CACHE;

switch (activeNetwork.getType()) {
    case (ConnectivityManager.TYPE_WIFI):
        PrefetchCacheSize = MAX_PREFETCH_CACHE;
        break;
    case (ConnectivityManager.TYPE_MOBILE): {
        switch (tm.getNetworkType()) {
            case (TelephonyManager.NETWORK_TYPE_LTE |
                    TelephonyManager.NETWORK_TYPE_HSPAP):
                PrefetchCacheSize * = 4;
                break;
            case (TelephonyManager.NETWORK_TYPE_EDGE |
                    TelephonyManager.NETWORK_TYPE_GPRS):
                PrefetchCacheSize / = 2;
                break;
            default: break;
        }
        break;
    }
    default: break;
}
```

3.3.3 在独立线程中执行网络连接

网络操作涉及不可预知的延迟,为了防止不良的用户体验,通常的做法是从 UI 中独立出线程去执行网络连接操作。AsyncTask 类提供了最简单地从 UI 线程中独立出一个新任务的方式。

AsyncTask 是抽象类,在 AsyncTask 中定义了 3 种泛型类型: Params、Progress 和 Result。

- Params 对应 doInBackground(Params...)方法的参数类型。而 new AsyncTask().execute(Params... params)传进来的就是 Params 数据,可以使用 execute(data)来传送一个数据,或者使用 execute(data1, data2, data3)传送多个数据。
- Progress 对应 onProgressUpdate(Progress...)的参数类型,显示后台任务执行的百分比。
- Result 对应 onPostExecute(Result)的参数类型,后台执行任务最终返回的结果,例如 String。

当以上的参数类型都不需要指明时,则使用 Void,注意不是 void。

AsyncTask 执行的每一步都对应一个回调方法,这些方法不应该由应用程序调用,开发者需要做的就是实现这些方法。首先子类化 AsyncTask,然后实现 AsyncTask 中定义的下面一个或几个方法。

- onPreExecute(): 执行预处理,它运行于 UI 线程,可以为后台任务做一些准备工作,例如绘制一个进度条控件。
- doInBackground(Params...): 在 onPreExecute()方法执行后马上自动执行,后台进程执行的具体计算在这里实现,是 AsyncTask 的关键,此方法必须重载。doInBackground()的返回值会传给 onPostExecute()。在 doInBackground()内的任何时刻,都可以调用 publishProgress()来执行 UI 线程中的 onProgressUpdate()方法以更新实时的任务进度。

  注意: doInBackground(Params...)中不能直接操作 UI。
- onProgressUpdate(Progress...): 运行于 UI 线程。如果在 doInBackground(Params...)中使用了 publishProgress(Progress...),就会触发这个方法。在这里可以对进度条控件根据进度值做出具体的响应。
- onPostExecute(Result): 运行于 UI 线程,相当于 Handler 处理 UI 的方式,在这里面可以使用在 doInBackground 中得到的结果处理 UI 操作。如果 Result 为 null 表明后台任务没有完成(被取消或者出现异常)。

可以看出,AsyncTask 的特点是在主线程之外运行任务,而回调方法是在主线程中执行,这就有效地避免了使用 Handler 带来的麻烦。

为了正确地使用 AsyncTask 类,必须遵守以下几条准则:

① AsyncTask 的实例必须在 UI 线程中创建。

② execute()方法必须在 UI 线程中调用。

③ 不要手动调用 onPreExecute()、onPostExecute(Result)、doInBackground(Params...)和 onProgressUpdate(Progress...)这几个方法。

④ AsyncTask 实例只能被执行一次,否则多次调用时将会出现异常。

⑤ 可以在任何时刻、任何线程内取消任务。

下面的代码片段演示了使用 AsyncTask 异步下载网页内容的方法。

```
public class AsyncTaskActivity extends Activity {
    private static final String DEBUG_TAG = "AsyncTask";
    private EditText urlText;
    private TextView textView;

    @Override
    public void onCreate(Bundle savedInstanceState) {
        super.onCreate(savedInstanceState);
        setContentView(R.layout.main);
        urlText = (EditText) findViewById(R.id.myUrl);
        textView = (TextView) findViewById(R.id.myText);
    }

    //当用户单击按钮则会调用 AsyncTask
    public void myClickHandler(View view) {
        String stringUrl = urlText.getText().toString();
        ConnectivityManager connMgr = (ConnectivityManager)
                getSystemService(Context.CONNECTIVITY_SERVICE);
        NetworkInfo networkInfo = connMgr.getActiveNetworkInfo();
        if (networkInfo != null && networkInfo.isConnected()) {
            new DownloadWebpageText().execute(stringUrl);
        } else {
            textView.setText("No network connection available.");
        }
    }

    private class DownloadWebpageText
            extends AsyncTask<String, Integer, String> {
        @Override
        protected String doInBackground(String... urls) {

            //参数来自 execute(),调用 params[0]得到 URL
            try {
                return downloadUrl(urls[0]);
            } catch (IOException e) {
                return "无法获取网页,URL 可能无效!Unable to retrieve web page. "
                        + "URL may be invalid.";
            }
        }
        //onPostExecute 显示 AsyncTask 结果
        @Override
        protected void onPostExecute(String result) {
            textView.setText(result);
        }
    }
    ...
}
```

这段代码中,myClickHandler()方法执行了 new DownloadWebpageTask().execute(stringUrl)。DownloadWebpageTask 类是 AsyncTask 的子类,DownloadWebpageTask 实现了 AsyncTask 的方法:doInbackground()会执行 downloadUrl()方法,downlaodUrl()方法将网页的 URL 地址作为参数,并获取和处理网页的内容,当它处理完这些操作,将会返回一个结果字符串。onPostExecute()接受返回字符串并显示在 UI 上。

3.4 习　　题

1. 编程实现以列表的形式列出所有可用的无线网络。
2. 编程实现从网络异步下载一幅图像并显示。

第 4 章 Socket 编程

Socket(套接字)是通信端点的抽象。Socket 提供了用于不同设备(通过网络相连)上运行的进程相互通信的接口,通过该接口可以在同一台设备上也可以在不同的设备上使用该接口和其他进程通信。

4.1 网络编程基础

本节主要介绍一些网络编程的基础知识。

4.1.1 TCP/IP 与网络通信

TCP/IP(Transport Control Protocol/Internet Protocol,传输控制协议/Internet 协议)起源于 20 世纪 60 年代末美国政府资助的一个分组交换网络研究项目,它是一个真正的开放协议。

TCP/IP 通信协议采用了四层的层级模型结构,每一层都调用它的下一层所提供的网络任务来完成自己的需求。TCP/IP 的每一层都是由一系列协议来定义的。这 4 层分别如下。

- 应用层(Application):应用层是个很广泛的概念,有一些基本相同的系统级 TCP/IP 应用以及应用协议,也有许多的企业商业应用和互联网应用。
- 传输层(Transport):传输层包括 UDP 和 TCP。UDP 几乎不对报文进行检查,而 TCP 提供传输保证。
- 网络层(Network):网络层协议由一系列协议组成,包括 ICMP、IGMP、RIP、OSPF、IP(v4,v6)等。
- 链路层(Link):又称为物理数据网络接口层,负责报文传输。

在现有的网络中,网络通信的方式主要有两种:TCP 方式和 UDP(User Datagram Protocol,用户数据报协议)方式。

- TCP 方式:传输控制协议提供的是面向连接、可靠的字节流服务。当客户和服务器彼此交换数据前,必须先在双方之间建立一个 TCP 连接,之后才能传输数据。TCP 提供超时重发、丢弃重复数据、检验数据、流量控制等功能,保证数据能从一端传到另一端。

  每个数据报的传输过程是:先建立链路,传输数据,然后清除链路。数据报不包含目的地址。收端和发端不但顺序一致,而且内容相同。它的可靠性高。
- UDP 方式:用户数据报协议是面向无连接的,每个数据报都有完整的源、目的地址及分组编号，各自在网络中独立传输，传输中不管其顺序，数据到达收端后再进行排序组装，遇有丢失、差错和失序等情况，通过请求重发来解决。它的效率比较高。

UDP 是一个简单的面向数据报的传输层协议。UDP 不提供可靠性，它只是把应用程序传给 IP 层的数据报发送出去，但是并不能保证它们能到达目的地。由于 UDP 在传输数据报前不用在客户和服务器之间建立一个连接，且没有超时重发等机制，故而传输速度很快。

因此，重要的数据一般使用 TCP 方式进行传输，而大量的非核心数据则都通过 UDP 方式进行传递，在一些程序中甚至这两种方式结合使用进行数据的传递。

4.1.2 C/S 模式与 B/S 模式

1. 客户/服务器模式

在 TCP/IP 网络应用中，通信的两个进程间相互作用的客户/服务器(Client/Server，C/S)模式的工作方式是：客户向服务器发出服务请求，服务器接收到请求后，提供相应的服务。客户/服务器模式的建立基于以下两点。

(1) 建立网络的起因是网络中软硬件资源、运算能力和信息不均等，需要共享，从而造就拥有众多资源的主机提供服务，资源较少的客户请求服务这一非对等状况。

(2) 网间进程通信完全是异步的，相互通信的进程间既不存在父子关系，又不共享内存缓冲区，因此需要一种机制为希望通信的进程间建立联系，为二者的数据交换提供同步，这就是基于客户/服务器模式的 TCP/IP。

在服务器端，服务器方要先启动，并根据请求提供以下的相应服务：

(1) 打开一通信通道并告知本地主机，它愿意在某一公认地址上的某个端口(如 FTP 的端口可能为 21)接收客户请求。

(2) 等待客户请求到达该端口。

(3) 接收到客户端的服务请求时，处理该请求并发送应答信号。接收到并发服务请求，要激活一个新进程来处理这个客户请求。新进程处理此客户请求，并不需要对其他请求做出应答。服务完成后，关闭此新进程与客户的通信链路并终止。

(4) 返回第(2)步，等待另一客户请求。

(5) 关闭服务器。

在客户端，其请求过程为：

(1) 打开一个通信通道，并连接到服务器所在主机的特定端口。

(2) 向服务器发服务请求报文，等待并接收应答。

(3) 继续提出请求。

(4) 请求结束后关闭通信通道并终止。

从上面所描述过程可知：客户与服务器进程的作用是非对称的，因此代码不同；服务器进程一般是先启动的。只要系统运行，该服务进程一直存在，直到正常或强迫终止。

2. 浏览器/服务器模式

浏览器/服务器(Browser/Server，B/S)模式是实现基于 Internet 即时通信的主要工作模式。

在 B/S 模式中，客户端运行浏览器软件。浏览器以超文本形式向 Web 服务器提出访问数据请求，请求的方式分为 POST 方式和 GET 方式。

- 对于 GET 请求，浏览器其实是一个 URL 请求，变量名和内容都包含在 URL 中，形式如 http://www.url.com/index.jsp? id=123。
- 对于 POST 请求，浏览器将生成一个数据报，用来将变量名和它们的内容捆绑在一起，并发送到服务器。

Web 服务器接受客户端请求后，如果是对静态页面的请求，就将静态页面发送给客户端。如果请求的内容需动态处理，请求将转交给动态处理程序，如 CGI、ASP、JSP 等，相应程序进行组件访问及数据库访问，将数据处理结果交给 Web 服务器。Web 服务器响应来自浏览器的请求，响应一般由状态行、某些响应头、一个空行和文档组成。客户端浏览器对服务器的响应进行解析，以友好的 Web 页面形式显示出来。

4.1.3 网络相关包

Android SDK 中包含了一些与网络有关的包，如表 4-1 所示。

表 4-1 Android SDK 网络包

包	描 述
java.net	提供与联网有关的类，包括流和数据报(datagram)sockets、Internet 协议和常见 HTTP 处理。该包是一个多功能网络资源
java.io	虽然没有提供显式的联网功能，但是该包中的类由其他 Java 包中提供的 socket 和连接使用。它们还用于与本地文件(在与网络进行交互时会经常出现)的交互
java.nio	包含表示特定数据类型的缓冲区的类。适合于两个基于 Java 语言的端点之间的通信
org.apache.*	包含许多为 HTTP 通信提供精确控制和功能的包。可以将 Apache 视为流行的开源 Web 服务器
android.net	除核心 java.net.* 类以外，包含额外的网络访问 socket。该包包括 URI 类，后者频繁用于 Android 应用程序开发，而不仅仅是传统的联网方面
android.net.http	包含处理 SSL 证书的类
android.net.wifi	包含在 Android 平台上管理有关 Wi-Fi(802.11 无线 Ethernet)所有方面的类
android.net.sip	包含 Andriod 平台上管理有关 SIP 协议，如建立和回应 Voip 的类
android.telephony.gsm	包含用于管理和发送 SMS 消息的类。一段时间后，可能会引入额外的包来为非 GSM 网络提供类似的功能，比如 CDMA 或 android.telephony.cdma 等网络
android.nfc	包含所有用来管理近场通信相关的功能类

其中的 java.net 包是标准的 Java 接口，提供与联网有关的类，包括流、数据报套接字、Internet 协议、常见 Http 处理等。该包是一个多功能网络资源。

java.net 包可以大致分为两个部分。

1. 低级 API

低级 API 主要用于处理以下抽象。

(1) 地址

地址也就是网络标识符，如 IP 地址。InetAddress 类是表示 IP(Internet 协议)地址的抽象。它拥有两个子类：

- 用于 IPv4 地址的 Inet4Address。

- 用于 IPv6 地址的 Inet6Address。

但是,在大多数情况下,不必直接处理子类,因为 InetAddress 抽象应该覆盖大多数必需的功能。

说明:并非所有系统都支持 IPv6 协议,而当 Java 网络连接堆栈尝试检测它并在可用时透明地使用它时,还可以利用系统属性禁用它。在 IPv6 不可用或被显式禁用的情况下,Inet6Address 对大多数网络连接操作都不再是有效参数。虽然可以保证在查找主机名时 java.net.InetAddress.getByName 之类的方法不返回 Inet6Address,但仍然可能通过传递字面值来创建此类对象。在此情况下,大多数方法在使用 Inet6Address 调用时都将抛出异常。

(2) 套接字

套接字也就是基本双向数据通信机制。套接字是在网络上建立机器之间通信连接的方法。java.net 包提供 4 种套接字:

- Socket 是 TCP 客户端 API,通常用于连接远程主机。
- ServerSocket 是 TCP 服务器 API,通常接受源于客户端套接字的连接。
- DatagramSocket 是 UDP 端点 API,用于发送和接收数据报。
- MulticastSocket 是 DatagramSocket 的子类,在处理多播组时使用。

使用 TCP 套接字的发送和接收操作需要借助 InputStream 和 OutputStream 来完成,这两者是通过 Socket.getInputStream()和 Socket.getOutputStream()方法获取的。

(3) 接口

用于描述网络接口。NetworkInterface 类提供 API 以浏览和查询本地机器的所有网络接口(例如,以太网连接或 PPP 端点)。只有通过该类才可以检查是否将所有本地接口都配置为支持 IPv6。

2. 高级 API

高级 API 主要用于处理以下抽象。

(1) URI

表示统一资源标识符。

(2) URL

表示统一资源定位符。

(3) 连接

表示到 URL 所指向资源的连接。

java.net 包中的许多类可以提供更加高级的抽象,允许方便地访问网络上的资源。这些类为:

- URI 是表示在 RFC 2396 中指定的统一资料标识符的类。顾名思义,它只是一个标识符,不直接提供访问资源的方法。
- URL 是表示统一资源定位符的类,它既是 URI 的旧式概念,又是访问资源的方法。
- URLConnection 是根据 URL 创建的,是用于访问 URL 所指向资源的通信连接。此抽象类将大多数工作委托给底层协议处理程序,如 http 或 ftp。
- HttpURLConnection 是 URLConnection 的子类,提供一些特定于 HTTP 协议的附加功能。

建议的用法是使用 URI 指定资源，然后在访问资源时将其转换为 URL。从该 URL 可以获取 URLConnection 以进行良好控制，也可以直接获取 InputStream。例如：

```
URI uri = new URI("http://www.baidu.com/");
URL url = uri.toURL();
InputStream in = url.openStream();
```

URL 和 URLConnection 都依赖于协议处理程序，所以协议处理程序必须存在，否则将抛出异常。此为与 URI 的主要不同点，URI 仅标识资源，所以不必访问协议处理程序。因此，尽管可能利用任何种类的协议方案(例如，myproto://myhost. mydomain/resource/)创建 URI，但类似的 URL 仍将试图实例化指定协议的处理程序，如果指定协议的处理程序不存在，则抛出异常。

默认情况下，协议处理程序从默认位置动态加载。但是，通过设置 java. protocol. handler. pkgs 系统属性也可能增加搜索路径。例如，如果将其设置为 myapp. protocols，则 URL 代码将首先尝试(对于 http 而言)加载 myapp. protocols. http. Handler，然后如果失败，则尝试从默认位置加载 http. Handler。

注意：处理程序类必须为抽象类 URLStreamHandler 的子类。

4.2 Socket 概述

Socket 是面向客户/服务器模型而设计的，针对客户和服务器程序提供不同的 Socket 系统调用，网络中的 Socket 数据传输可以理解为一种特殊的 I/O。

4.2.1 什么是 Socket 通信

在网络出现之前的单机系统时代，更多的是进程之间的通信。由于每个进程有自己的地址空间，为了保证两个进程的通信互不干扰又协调工作，操作系统提供了相应的设施。如 UNIX 系统中的管道、命名管道和软中断信号。而在网络中，两个不同计算机中的进程需要通信首先要解决的是进程识别的问题。同一主机上，不同进程可以用进程号作为唯一标识，但是在网络环境里不同的主机上完全可以用同一进程号，所以用这种方法来区别进程是不可行的。另外，操作系统支持的网络协议很多，不同协议的工作方式是不一样的，包括网络的地址格式也不同。因此，还需要考虑不同网络协议的识别问题。

为了解决这个问题，提出了 Socket 这个概念。Socket 又称为"套接字"，用于描述网络地址与端口[1]，它是一个通信的接口，是应用层与 TCP/IP 协议族通信的中间的软件抽象层，位于运输层和网络层之上，又位于应用层之下，作为一个抽象层存在。图 4-1 描述了

① 在 Socket 通信中 TCP 以及 UDP 都是采用 16 位的端口号来识别应用程序，服务器一般都是通过端口号来识别的。客户端通常对它使用的端口并不关心，只需保证它是唯一的就可以了，客户端的端口号又称为临时端口号(即存在时间很短)。这是因为它通常在客户运行该客户端程序时才存在。端口号最大范围是 0～65535。其中 1～1023 的端口号是已经被预先占有的服务端口号。大多数的临时端口分配范围在 1024～5000。大于 5000 的端口号是为其他服务器预留的(Internet 上并不常用)。

Socket 的作用和所处的位置。

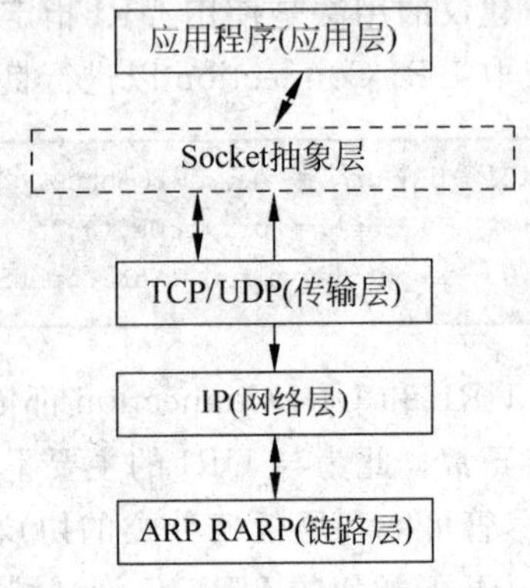

图 4-1　Socket 在 TCP/IP 中的位置

有时多个应用程序可能同时需要向同一个接口发送数据。为了区别不同的应用程序进程和连接，许多计算机操作系统为应用程序与 TCP/IP 协议交互提供了称为 Socket 的接口，或者说套接字 Socket 是 TCP 的应用编程接口 API，通过它应用层就可以访问 TCP 提供的服务。

区分不同应用程序进程间的网络通信和连接，主要有 3 个参数：通信的目的 IP 地址、使用的传输层协议(TCP 或 UDP)和使用的端口号。通过将这 3 个参数结合起来，与一个 Socket 绑定，应用层就可以和传输层通过 Socket 接口，区分来自不同应用程序进程或网络连接的通信，实现数据传输的并发服务。

4.2.2　Socket 通信的基本步骤

Socket 通信的过程可以分为三个步骤：服务器监听，客户端请求，连接确认。

(1) 服务器监听：是服务器端 Socke 处于等待连接的状态，实时监控网络状态。

(2) 客户端请求：是指由客户端的 Socket 提出连接请求，要连接的目标是服务器端的 Socket。为此，客户端的 Socket 必须首先描述它要连接的服务器的 Socket，指出服务器端 Socket 的地址和端口号，然后向服务器端 Socket 提出连接请求。

(3) 连接确认：是指当服务器端 Socket 监听到客户端 Socket 的连接请求，它就响应客户端 Socket 的请求，建立一个新的线程，把服务器端 Socket 的描述发给客户端，一旦客户端确认了此描述，连接就建立好了。而服务器端 Socket 继续处于监听状态，继续接收其他客户端 Socket 的连接请求。

因此，Socket 可以看成在两个程序进行通信连接中的一个端点，是连接应用程序和网络驱动程序的桥梁，Socket 在应用程序中创建，通过绑定与网络驱动建立关系。此后，应用程序发送给 Socket 的数据，由 Socket 通过网络驱动程序向网络上发送出去。计算机从网络上收到与该 Socket 绑定 IP 地址和端口号相关的数据后，由网络驱动程序交给 Socket，应用程序便可从该 Socket 中提取接收到的数据，网络应用程序就是这样通过 Socket 进行数据的发送与接收的。

1. Socket 通信连接过程

Socket 连接建立是基于 TCP 的连接建立过程，TCP 的连接需要通过三次握手报文来完成。在 TCP 协议中，是通过 3 个 TCP 格式的 IP 数据报来实现的。

TCP 格式的 IP 数据报中包含着 TCP 首部，TCP 首部信息中包含着对每一个数据报具体内容的描述。这些首部包括以下内容。

- SYN：同步序号用来发起一个连接。当发起连接时，SYN 就会被设置成 1。因为 TCP 协议要求数据传送是可靠的，其实现方式就是对传输数据的每一个字节按顺序编号。初始序列号 ISN 并非从 0 开始，而是一个随时间周而复始变化的 32 位无符号整数。发起连接后，假设发起连接一方的 ISN 为 n，因为 SYN 会在数据部分添加一个字节表示这是一个新连接的开始，所以这时字节序号就成了 $n+1$。

- ACK：确认序号有效。TCP 协议要求自动检验数据的可靠性，实现方式就是检验字节序号是否正确地衔接。假如接收数据的一方序号已经是 m，那么其返回给发送方确认有效的序号就是 $m+1$。一旦连接，ACK 始终设置为 1，即表示序号有效，并且在所有数据报中总是存在。

Socket 连接建立时的三次握手过程是：开始建立 TCP 连接时需要发送同步 SYN 报文，然后等待确认报文 SYN＋ACK，最后再发送确认报文 ACK。具体过程如下。

- 客户端发送 SYN 包(SYN=n)到服务器，并进入 SYN_SEND 状态，等待服务器确认。
- 服务器端收到第一次握手请求的数据报后开始构建反馈的数据报。反馈数据报包括两个部分(SYN＋ACK)：第一部分是将连接请求的序号反馈回去，因为 SYN 本身占了一个字节，所以反馈回去的序号就是 ACK($n+1$)；第二部分是向客户端发起 SYN 连接请求，包含这个新连接的 ISN，设其值为 m。
- 客户端收到服务器的 SYN＋ACK 包，向服务器发送确认包 ACK($m+1$)，因为 SYN 占了一个字节，所以反馈给服务器端的序号是 $m+1$。

握手过程中传送的包里不包含数据，三次握手完毕后，客户端与服务器才正式开始传送数据。理想状态下，TCP 连接一旦建立，在通信双方中的任何一方主动关闭连接之前，TCP 连接都将被一直保持下去。

2. Socket 通信断开连接过程

TCP 的断开连接需要通过四次握手报文来完成，在 TCP 协议中，是通过 TCP 格式的 IP 数据报来实现的。

TCP 格式的 IP 数据报中包含着 TCP 首部。在断开连接时，TCP 首部信息中包含着对每一个数据报具体内容的描述。

FIN 表示发送端完成发送。与 SYN 类似，FIN 也会在数据部分占用一个字节，表示这是一个结束符号。由于 TCP 连接是全双工的，因此每个方向都必须单独进行关闭。原则是当一方完成它的数据发送任务后，就能发送一个 FIN 来终止这个方向的连接。收到一个 FIN 只意味着这一方向上没有数据流动，一个 TCP 连接在收到一个 FIN 后仍能发送数据。首先进行关闭的一方将执行主动关闭，而另一方执行被动关闭。

Socket 开始断开 TCP 连接时需要发送 FIN 报文，等待确认报文 ACK，然后由接收方再次发给发送方一个断开连接 FIN 报文，等待原发送方确认报文 ACK，这样就断开了连接。具体过程如下：

① TCP 客户端发送一个 FIN，用来关闭客户到服务器的数据传送。

② 服务器收到这个 FIN，它发回一个 ACK，确认序号为收到的序号加 1。与 SYN 一样，一个 FIN 将占用一个序号。

③ 服务器关闭客户端的连接，发送一个 FIN 给客户端。

④ 客户端发回 ACK 报文进行确认，并将确认序号设置为收到序号加 1。

4.3　Android 中的 Socket 编程

本节介绍 Android 中与 Socket 编程密切相关的类，并通过一个案例介绍 Android 中进行 Socket 编程的基本方法和步骤。

4.3.1 Socket 相关类

1. InetAddress

IP 地址是 IP 使用的 32 位(IPv4)或者 128 位(IPv6)位无符号数字,它是传输层协议 TCP、UDP 的基础。java.net.InetAddress 是 Java 对 IP 地址的封装,在 java.net 中有许多类都使用到了 InetAddress,包括 ServerSocket、Socket、DatagramSocket 等。

InetAddress 的实例对象包含以数字形式保存的 IP 地址,同时还可能包含主机名(如果使用主机名来获取 InetAddress 的实例,或者使用数字来构造,并且启用了反向主机名解析的功能)。InetAddress 类提供了将主机名解析为 IP 地址(或反之)的方法。

InetAddress 对域名进行解析是使用本地机器配置或者网络命名服务(如域名系统(Domain Name System,DNS)和网络信息服务(Network Information Service,NIS))来实现。对于 DNS 来说,本地需要向 DNS 服务器发送查询的请求,然后服务器根据一系列的操作,返回对应的 IP 地址。为了提高效率,通常本地会缓存一些主机名与 IP 地址的映射,这样访问相同的地址,就不需要重复发送 DNS 请求了。在 java.net.InetAddress 类中同样采用了这种策略。默认情况下,会缓存一段有限时间的映射。对于主机名解析不成功的结果,会缓存非常短的时间(10 秒)来提高性能。

InetAddress 的构造方法不是公开的(因此,不能直接创建 InetAddress 对象,如 InetAddress ia = new InetAddress ()是错误的),所以需要通过它提供的静态方法(即工厂模式)来获取,有以下的方法。

- static InetAddress[] getAllByName(String host):在给定主机名的情况下,根据系统上配置的名称服务返回其 IP 地址所组成的数组。
- static InetAddress getByAddress(byte[] addr):在给定原始 IP 地址的情况下,返回 InetAddress 对象。
- static InetAddress getByAddress(String host,byte[] addr):根据提供的主机名和 IP 地址创建 InetAddress。
- static InetAddress getByName(String host):返回一个 InetAddress 对象,该对象包含了一个与 host 参数指定的域名或 IP 地址,对于指定的主机如果没有 IP 地址存在,那么方法将抛出一个 UnknownHostException 异常对象。例如:

```
InetAddress ia = InetAddress.getByName("www.baidu.com");
InetAddress ia = InetAddress.getByName("111.13.100.92");
```

- static InetAddress getLocalHost():返回一个 InetAddress 对象,这个对象包含了本地机域名和 IP 地址。

在这些静态方法中,最为常用的应该是 getByName(String host)方法,只需要传入目标主机的名字,InetAddress 就会尝试连接 DNS 服务器并且获取 IP 地址。

InetAddress 对象可以映射地址,也可以通过其获取相关属性。常用的方法如下。

- String getHostName():用于获得 InetAddress 映射地址对象中的域名或者主机名。

```
String dominName = inetAddress.getHostName();
```

- String getHostAddress()：返回一个 InetAddress 映射地址对象的 IP 地址信息。

```
String ip = inetAddress.getHostAdress();
```

注意：这些方法可能会抛出的异常。如果安全管理器不允许访问 DNS 服务器或禁止网络连接，SecurityException 会抛出。如果找不到对应主机的 IP 地址，或者发生其他网络 I/O 错误，这些方法会抛出 UnknowHostException。

下面的方法演示了获取本地 IP 地址的方法。

```
public static String getLocalHostIp() {

    try {
        Enumeration<NetworkInterface> en = NetworkInterface
                .getNetworkInterfaces();

        while (en.hasMoreElements()) {
            NetworkInterface nif = en.nextElement();
            Enumeration<InetAddress> inet = nif.getInetAddresses();

            while (inet.hasMoreElements()) {
                InetAddress ip = inet.nextElement();
                if (!ip.isLoopbackAddress()
                        && InetAddressUtils.isIPv4Address(ip
                                .getHostAddress())) {
                    return ip.getHostAddress();
                }
            }

        }
    } catch (SocketException e) {
        Log.e("InetAddressActivity", "获取本地 IP 地址失败");
        e.printStackTrace();
    }

    return IP_DEFAULT;          //"0.0.0.0"
}
```

2. Socket

java.net.Socket 类用于描述 IP 地址和端口，是一个通信链的句柄，可以实现客户端套接字。应用程序通过 Socket 向网络发出请求或者应答网络请求。

在网络连接成功时，应用程序两端都会产生一个 Socket 实例，通过操作这个实例，完成所需的会话。对于一个网络连接来说，套接字是平等的，并没有差别，不会因为在服务器端或在客户端而产生不同级别。不管是 Socket 还是 ServerSocket，它们的工作都是通过 SocketImpl 类及其子类完成的。

(1) 常用构造方法

java.net.Socket 继承自 java.lang.Object。下面介绍两个常见的构造方法。

- Socket(InetAddress address, int port):创建一个流套接字并将其连接到指定 IP 地址的指定端口号。
- Socket(String host, int port):创建一个流套接字并将其连接到指定主机上的指定端口号。

两个构造方法都创建了一个基于 Socket 的连接服务器端流套接字的流套接字。对于第一个方法中的 InetAddress 子类对象,通过 address 参数获得服务器主机的 IP 地址。对于第二个方法,host 参数包被分配到 InetAddress 对象中,如果没有 IP 地址与 host 参数相一致,那么将抛出 UnknownHostException 异常对象。两个方法都通过 port 参数获得服务器的端口号。假设已经建立了连接,网络 API 将在客户端基于 Socket 的流套接字中捆绑客户程序的 IP 地址和任意一个端口号,否则两个方法都会抛出一个 IOException 对象。

如果应用程序已指定套接字工厂,则调用该工厂的 createSocketImpl()方法来创建实际套接字实现。否则创建"普通"套接字。如果有安全管理器,则使用主机地址和 port 作为参数调用其 checkConnect()方法。这可能会导致 SecurityException 异常。

(2) 常用方法

下面介绍几个使用最频繁的方法。

- InputStream getInputStream():返回此套接字的输入流。因关闭了返回的 InputStream,所以将关闭关联套接字。

  如果此套接字具有关联的通道,则所得的输入流会将其所有操作委托给通道。如果通道为非阻塞模式,则输入流的 read()操作将抛出 IllegalBlockingModeException。在非正常条件下,底层连接可能被远程主机或网络软件中断(例如,TCP 连接情况下的连接重置)。当网络软件检测到中断的连接时,将对返回的输入流应用以下操作:

  - 网络软件可能丢弃经过套接字缓冲的字节。网络软件没有丢弃的字节可以使用 read()读取。
  - 如果没有任何字节在套接字上缓冲,或者 read()已经消耗了所有缓冲的字节,则对 read()的所有后续调用都将抛出 IOException。
  - 如果没有任何字节在套接字上缓冲,并且没有使用 close()关闭套接字,则 available()将返回 0。

- OutputStream getOutputStream():返回此套接字的输出流。因关闭了返回的 OutputStream,所以将关闭关联套接字。

  如果此套接字具有关联的通道,则得到的输出流会将其所有操作委托给通道。如果通道为非阻塞模式,则输出流的 write()操作将抛出 IllegalBlockingModeException 异常。

- void connect(SocketAddress endpoint):将此套接字连接到服务器。isConnected()方法返回套接字的连接状态。
- void connect(SocketAddress endpoint, int timeout):将此套接字连接到服务器,并指定一个超时值。timeout 等于 0 表示无限超时。在建立连接或者发生错误之前,连接一直处于阻塞状态。
- void bind(SocketAddress bindpoint):将套接字绑定到本地地址。如果地址为

null，则系统将挑选一个临时端口和一个有效本地地址来绑定套接字。isBound()方法返回套接字的绑定状态。

- void close()：关闭此套接字。isClosed()方法返回套接字的关闭状态。
- InetAddress getInetAddress()：返回套接字连接的地址。
- InetAddress getLocalAddress()：获取套接字绑定的本地地址。
- SocketAddress getLocalSocketAddress()：返回此套接字绑定的端点的地址。如果尚未绑定，则返回 null。
- SocketAddress getRemoteSocketAddress()：返回此套接字连接的端点的地址，如果未连接，则返回 null。
- int getPort()：返回此套接字连接到的远程端口。
- int getLocalPort()：返回此套接字绑定到的本地端口。

3. ServerSocket

在 Client/Server 通信模式中，服务器端需要创建监听特定端口的 ServerSocket，java.net.ServerSocket 所封装的底层操作会作为服务的提供端来监听某一个网络端口，并等待客户端的连接请求。

(1) 构造方法

java.net.ServerSocket 继承自 java.lang.Object，构造方法包括以下几种。

① ServerSocket()：创建非绑定服务器套接字。

ServerSocket 不带参数的默认构造方法创建的 ServerSocket 不与任何端口绑定，接下来还需要通过 bind()方法与特定端口绑定。这个默认构造方法的用途是，允许服务器在绑定到特定端口之前先设置 ServerSocket 的一些选项。因为一旦服务器与特定端口绑定，有些选项就不能再改变了。

② ServerSocket(int port)：创建绑定到特定端口的服务器套接字。

如果 port 为 0，则在所有空闲端口上创建套接字。传入连接指示(对连接的请求)的最大队列长度被设置为 50。管理客户连接请求的任务是由操作系统来完成的，如果队列满时收到连接指示，则拒绝该连接。

对于客户进程，如果它发出的连接请求被加入服务器的队列中，就意味着客户与服务器的连接建立成功，客户进程从 Socket 构造方法中正常返回。如果客户进程发出的连接请求被服务器拒绝，Socket 构造方法就会抛出 ConnectionException。

如果运行时无法绑定到 port 端口，会抛出 BindException 异常。BindException 一般是由两种原因造成的：端口已经被其他服务器进程占用；在某些操作系统中，如果没有以超级用户的身份来运行服务器程序，那么操作系统不允许服务器绑定到 1～1023 的端口。

③ ServerSocket(int port, int backlog)：利用指定的 backlog 创建服务器套接字并将其绑定到指定的本地端口号，backlog 将覆盖操作系统限定的队列的最大长度。

传入连接指示(对连接的请求)的最大队列长度被设置为 backlog 参数。如果队列满时收到连接指示，则拒绝该连接。backlog 参数必须是大于 0 的正值。如果传递的值等于或小于 0，则使用默认值。

对于 ServerSocket(int port)和 ServerSocket(int port, int backlog)方法，如果应用程序已指定服务器套接字工厂，则调用该工厂的 createSocketImpl()方法来创建实际套接字实现,否则创建"普通"套接字。如果存在安全管理器,则首先使用 port 参数作为参数调用其 checkListen()方法,以确保允许该操作,这可能会导致 SecurityException 异常。

④ ServerSocket(int port, int backlog, InetAddress bindAddr):使用指定的端口、侦听 backlog 和要绑定到的本地 IP 地址创建服务器。

bindAddr 参数可以在 ServerSocket 的多宿主主机上使用,ServerSocket 仅接受对其地址之一的连接请求。如果 bindAddr 为 null,则默认接受所有本地地址上的连接。端口必须为 0～65535(包括两者)。

(2) 常用方法

下面介绍几个使用最频繁的方法。

- Socket accept():从连接请求队列中取出一个客户的连接请求,然后创建与客户连接的 Socket 对象,并将它返回。如果队列中没有连接请求,accept()方法就会一直等待,直到接收到了连接请求才返回。只有当服务器进程通过 ServerSocket 的 accept()方法从队列中取出连接请求,使队列腾出空位时,队列才能继续加入新的连接请求。
- void bind(SocketAddress endpoint):将 ServerSocket 绑定到特定地址(IP 地址和端口号)。如果地址为 null,则系统将挑选一个临时端口和一个有效本地地址来绑定套接字。isBound()方法返回 ServerSocket 的绑定状态。只要 ServerSocket 已经与一个端口绑定,即使它已经被关闭,isBound()方法也会返回 true。
- void bind(SocketAddress endpoint, int backlog):将 ServerSocket 绑定到特定地址(IP 地址和端口号)。backlog 参数必须是大于 0 的正值。如果传递的值等于或小于 0,则使用默认值。

说明:bind()方法中要有表达地址和端口的重要参数,类型为 SocketAddress,而应用中使用子类 InetSocketAddress。常用的实现方式有以下几种。

```
InetSocketAddress ia = new InetSocketAddress(int port);
InetSocketAddress ia = new InetSocketAddress(InetAddress addr, int port);
InetSocketAddress ia = new InetSocketAddress(String hostname, int port);
```

- void close():使服务器释放占用的端口,并且断开与所有客户的连接。在 accept()中所有当前阻塞的线程都将会抛出 SocketException。如果此套接字有一个与之关联的通道,则关闭该通道。isClosed()方法返回 ServerSocket 的关闭状态。只有执行了 ServerSocket 的 close()方法,isClosed()方法才返回 true;否则,即使 ServerSocket 还没有和特定端口绑定,isClosed()方法也会返回 false。
- InetAddress getInetAddress():返回此服务器套接字的本地地址。如果套接字是未绑定的,则返回 null。
- SocketAddress getLocalSocketAddress():返回此套接字绑定的端点的地址,如果尚未绑定,则返回 null。
- int getLocalPort():返回此套接字在其上侦听的端口。如果尚未绑定套接字,则返回 −1。

- protected void implAccept(Socket s)：ServerSocket 的子类使用此方法重写 accept()以返回它们自己的套接字子类。

4.3.2 实现 Socket 通信

1. 客户端编程

客户端是 Socket 网络编程中首先发起连接的程序，客户端的编程主要由三个步骤实现。

(1) 建立网络连接

客户端 Socket 编程的第一步是建立网络连接。在建立网络连接时需要指定连接到的服务器的 IP 地址和端口号，建立完成以后，会形成一条虚拟的连接，后续的操作就可以通过该连接实现数据交换了。

(2) 交换数据

连接建立以后，交换数据严格按照请求响应模型进行，由客户端发送一个请求数据到服务器，服务器反馈一个响应数据给客户端，如果客户端不发送请求则服务器端就不响应。根据逻辑需要，可以多次交换数据，但是必须遵循请求响应模型。

(3) 关闭网络连接

在数据交换完成以后，关闭网络连接，释放程序占用的端口、内存等系统资源，结束网络编程。

下面通过一个实例演示客户端 Socket 编程的一般步骤。

① 用服务器的 IP 地址和端口号实例化 Socket 对象。

在 ClientActivity 的 onCreate()生命周期方法中，调用如下代码初始化一个客户端线程。

提示：Android 为了防止应用的 ANR(aplication Not Response)异常，在 Android Honeycomb 及之后的 API 版本中，对 UI 线程进行联网操作是不允许的。如果有网络操作，会抛出 NetworkOnMainThreadException 的异常。

```
if(_serverPort > 0 && _serverAddress != null){
    this._clientThread = new Thread(new ClientThread());
    this._clientThread.start();
    try{
        this._clientThread.join();
    }catch (Exception e){
        e.printStackTrace();
    }
}
```

其中，ClientThread 的代码如下：

```
final class ClientThread implements Runnable {

    @Override
    public void run() {

```

```
06          try {
07              InetAddress serverAddr = InetAddress.getByName(_serverAddress);
08              _socket = new Socket(serverAddr, _serverPort);
09          } catch (UnknownHostException e1) {
10              e1.printStackTrace();
11          } catch (IOException e1) {
12              e1.printStackTrace();
13          }
14      }
15
16  }
```

其中的 07～08 行实例化一个 Socket 对象。

② 调用 connect()方法，连接到服务器上。

③ 将发送到服务器的 I/O 流填充到 I/O 对象里，比如 BufferedReader()/PrintWriter()。

④ 利用 Socket 提供的 getInputStream()和 getOutputStream()方法，通过 I/O 流对象，向服务器发送数据流。

例如，在 ClientActivity 中，单击界面中的"发送"按钮后，调用 sendToServer()方法向服务端发送数据，代码如下：

```
public void sendToServer(String datas){
    try {
        PrintWriter out = new PrintWriter(new BufferedWriter(
                new OutputStreamWriter(_socket.getOutputStream())),
                true);
        out.println(datas);
    } catch (UnknownHostException e) {
        e.printStackTrace();
    } catch (IOException e) {
        e.printStackTrace();
    } catch (Exception e) {
        e.printStackTrace();
    }
}
```

⑤ 通信完成后，关闭打开的 I/O 对象和 Socket。

例如，下面代码的 05 行将客户端线程置为 Interrupted 状态。

```
01  @Override
02  protected void onStop() {
03      super.onStop();
04      try{
05          this._clientThread.interrupt();
06      }catch (Exception e){
07          e.printStackTrace();
08      }
09  }
```

当一个 Socket 对象被关闭，就不能再通过它的输入流和输出流进行 I/O 操作，否则会导致 IOException 异常。为了确保关闭 Socket 的操作总是被执行，强烈建议把这个操作放在 finally 代码块中。例如：

```
try {
    ...
} catch (UnknownHostException e) {
    e.printStackTrace();
} catch (IOException e) {
    e.printStackTrace();
} finally {
    try {
        client.close();
    } catch (Exception e) {
        e.printStackTrace();
    }
}
```

2. 服务端编程

服务端是在 Socket 网络编程中被动等待连接的程序，服务端的编程步骤和客户端不同，是由以下四个步骤实现的。

(1) 监听端口

服务端属于被动等待连接，所以服务端启动以后，不需要发起连接，而只需要监听本地某个固定端口即可。这个端口就是服务端开放给客户端的端口，服务端程序运行的本地 IP 地址就是服务端程序的 IP 地址。

(2) 获得连接

当客户端连接到服务端时，服务端就可以获得一个连接，这个连接包含客户端的信息，例如客户端 IP 地址等，服务端和客户端也通过该连接进行数据交换。一般在服务端编程中，当获得连接时，需要开启专门的线程处理该连接，每个连接都由独立的线程实现。

(3) 交换数据

服务端通过获得的连接进行数据交换。服务端的数据交换步骤是首先接收客户端发送过来的数据，然后进行逻辑处理，再把处理以后的结果数据发送给客户端。简单来说，就是先接收再发送，这个和客户端的数据交换数序不同。当然，服务端的数据交换也是可以多次进行的。在数据交换完成以后，关闭和客户端的连接。

(4) 关闭连接

当服务器程序关闭时，需要关闭服务端，通过关闭服务端使得服务器监听的端口以及占用的内存可以释放出来，实现连接的关闭。

下面通过一个实例演示服务端 ServerSocket 编程的一般步骤。

① 在服务器上用一个端口来实例化一个 ServerSocket 对象。此时，服务器就可以用这个端口时刻监听从客户端发来的连接请求。

在 ServerActivity 的 onCreate()生命周期方法中，调用如下代码初始化一个服务端线程：

```
if(_port > 0){
    _serverThread = new Thread(new ServerRunnable());
    _serverThread.start();
}
```

其中,ServerRunnable 代码如下:

```
01 final class ServerRunnable implements Runnable{
02
03     @Override
04     public void run() {
05         Socket _socket = null;
06
07         try{
08             _serverSocket = new ServerSocket(_port);
09         }catch (IOException e){
10             e.printStackTrace();
11             return;
12         }
13
14         while (!Thread.currentThread().isInterrupted()) {
15
16             try {
17                 _socket = _serverSocket.accept();
18                 CommunicationThread commThread =
19                         new CommunicationThread(_socket);
20                 new Thread(commThread).start();
21
22             } catch (IOException e) {
23                 e.printStackTrace();
24             }
25         }
26     }
27 }
```

② 调用 ServerSocket 的 accept()方法开始监听从端口发来的连接请求。

③ 利用 accept()方法返回的客户端的 Socket 对象进行读写 I/O 的操作。如上面代码的第 17 行。其中负责数据交互的 CommunicationThread 代码如下:

```
final class CommunicationThread implements Runnable {

    private Socket clientSocket;
    private BufferedReader input;

    public CommunicationThread(Socket clientSocket) {

        this.clientSocket = clientSocket;

```

```
10            try {
11                this.input = new BufferedReader(new InputStreamReader(
12                        this.clientSocket.getInputStream()));
13            } catch (IOException e) {
14                e.printStackTrace();
15            }
16        }
17
18        public void run() {
19
20            while (!Thread.currentThread().isInterrupted()) {
21                try {
22                    final String read = input.readLine();
23
24                    runOnUiThread(new Runnable() {
25                        @Override
26                        public void run() {
27                            onDataReceive(read);
28                        }
29                    });
30                } catch (IOException e) {
31                    e.printStackTrace();
32                }
33            }
34        }
35
36    }
```

其中,第 27 行的 onDataReceive()方法负责将从客户端收到的数据显示在用户界面上。

④ 通信完成后,关闭打开的 I/O 流和 ServerSocket 对象。例如:

```
@Override
protected void onStop() {
    super.onStop();
    try{
        _serverSocket.close();
    }catch (Exception e){
        e.printStackTrace();
    }
    try {
        _serverThread.interrupt();
    }catch (Exception e){
        e.printStackTrace();
    }
}
```

下面介绍模拟器之间的 Socket 通信。

两个模拟器之间 Socket 通信设置的基本步骤如下。

(1) 获取模拟器的 IP

① 启动一个模拟器。

② 启动 Windows 的命令行窗口,进入 Android sdk/platform-tools 目录。

③ 输入 adb shell,进入 Shell 模式。

④ 输入 getprop 命令将会显示系统的各项属性,其中包括模拟器的 DNS 地址,例如:

```
Microsoft Windows [版本 6.1.7601]
版权所有 (c) 2009 Microsoft Corporation.保留所有权利。
D:\android-sdk-windows>cd platform-tools
D:\android-sdk-windows\platform-tools>adb shell
root@generic_x86:/ # getprop
getprop
...
[net.bt.name]: [Android]
[net.change]: [net.dns1]
[net.dns1]: [10.0.2.3]
[net.eth0.dns1]: [10.0.2.3]
[net.eth0.gw]: [10.0.2.2]
[net.gprs.local-ip]: [10.0.2.15]
[net.hostname]: [android-91627a890f7bd0a5]
```

可以发现,在对两个模拟器进行获取时,IP 地址完全一样,都是 10.0.2.15,DNS 都是 10.0.2.3,所以要实现两个模拟器之间的通信,使用模拟器的 IP 地址是办不到的。

注意: net.eth0.gw 地址 10.0.2.2 等同于 PC 本机的 IP 地址 127.0.0.1。

(2) 设置模拟器的 IP

① 进入 Shell 模式。

② 输入 setprop net.dns1 192.168.0.1,即可把 DNS 修改成 PC 的 DNS。

③ 输入 setprop net.gprs.local-ip 192.168.0.108,即可把 IP 地址修改为与 PC 在一个网段。

这样模拟器的 IP 地址就设置完成了。

(3) 重定向模拟器(以 emulator-5554 作为服务端)

① 启动 Windows 的命令行窗口。

② 输入 telnet localhost 5554,连接到模拟器 5554,显示如下信息:

```
Trying ::1...
Trying 127.0.0.1...
Connected to localhost.
Escape character is '^]'.
Android Console: type 'help' for a list of commands
OK
```

③ 成功连接后,继续执行"redir add tcp:5000:6000"命令,将 PC 端口 5000 绑定到模拟器 5554 的端口 6000 上。

此时模拟器 5556 通过向 PC 的端口 5000(即地址 10.0.2.2:5000)发送 TCP/UDP 数

据报，即可与模拟器 5554 通信。

4.4 UDP 编程与 NIO 编程

本节简单介绍关于 UDP 编程和 NIO 编程的基础知识。

4.4.1 UDP 编程

4.3 节介绍的 Socket 编程是基于 TCP 协议的，本节介绍使用 UDP 实现 DatagramSocket 编程的方法。

UDP(User Datagram Protocol，用户数据报协议)是 OSI 参考模型中一种无连接的传输层协议，提供面向事务的简单不可靠信息传送服务。在网络中它与 TCP 协议一样用于处理数据报。UDP 有不提供数据报分组、组装和不能对数据报进行排序的缺点，也就是说，当报文发送之后，是无法得知其是否安全完整到达的。

类似网络视频会议系统等众多的客户/服务器模式的网络应用都需要使用 UDP 协议。UDP 协议的主要作用是将网络数据流量压缩成数据报的形式。一个典型的数据报就是一个二进制数据的传输单位。每一个数据报的前 8 个字节用来包含报头信息，剩余字节则用来包含具体的传输数据。

实现 UDP 编程的方法如下。

1. 接收方编程

接收方编程的一般步骤如下。

① 创建一个 DatagramSocket 对象，并指定监听的端口号。

DatagramSocket 表示接收或发送数据报的套接字，位于 java.net 包中。DatagramSocket 的构造方法有以下几种。

- DatagramSocket()：通常用于客户端编程，它用本地任何一个可用的端口创建一个套接字，这个端口号是由系统随机产生的。如果构造不成功，则触发 SocketException 异常。使用方法如下：

```
try{
    DatagramSocket datas = new DatagramSocket();
} catch(SocketException e){
}
```

- DatagramSocket(int port)：用一个指定的端口号 port 创建一个套接字。
- DatagramSocket(int port, InetAddress localAddr)：当一台设备拥有多于一个 IP 地址的时候，由它创建的实例仅仅接收来自 LocalAddr 的报文。

DatagramSocket 提供的常用方法包括以下几个。

- pubic void close()：关闭 DatagramSocket。在应用程序退出的时候，通常会主动释放资源，关闭 Socket，但是由于异常退出可能造成资源无法回收。所以，应该在程序完成时，主动使用此方法关闭 Socket，或在捕获到异常抛出后关闭 Sock。

- public int getLocalPort()：返回本地套接字正在监听的端口号。
- public void receive(DatagramPacket p)：从网络上接收数据报并将其存储在 DatagramPacket 对象 p 中。p 中的数据缓冲区必须足够大，receive()把尽可能多的数据存放在 p 中，如果装不下，就把其余的部分丢弃。接收数据出错时会抛出 IOException 异常。
- publicvoid Send(DatagramPacket p)：发送数据报，出错时会发生 IOException 异常。
- setSoTimeout(int timeout)：设置超时时间，单位为毫秒。

② 创建一个 byte 数组用于接收数据。如：

```
byte data[] = new byte[1024];
```

③ 创建一个空的 DatagramPackage 对象。

DatagramPackage 表示存放数据的数据报，位于 java. net 包中。DatagramPackage 主要用于将 byte 数组、目标地址、目标端口等数据包装成报文或者将报文拆卸成 byte 数组。

DatagramPackage 接收数据报的常用构造方法有两种：

- DatagramPacket(byte ibuft[],int ilength)
- DatagramPacket(byte ibuft[],int offset,int ilength)

ibuf[]为接受数据报的存储数据的缓冲区长度，ilength 为从传递过来的数据报中读取的字节数。当采用第一种构造方法时，接收到的数据从 ibuft[0]开始存放，直到整个数据报接收完毕或者将 ilength 的字节写入 ibuft 为止。采用第二种构造方法时，接收到的数据从 ibuft[offset]开始存放。如果数据报长度超出了 ilength，则触发 IllegalArgument-Exception(该异常是 RuntimeException，一般不需要用户单独写代码捕获它)。

DatagramPackage 发送数据报的常用构造方法有两种：

- DatagramPacket(byt ibuf[],int ilength,InetAddrss iaddr,int port)
- public DatagramPacket(byt ibuf[],int offset, int ilength,InetAddrss iaddr,int port)

iaddr 为数据报要传递到的目标地址，port 为目标地址的程序接受数据报的端口号。ibuf[]为要发送数据的存储区，以 ibuf 数组的 offset 位置开始填充数据报的 ilength 字节。如果没有 offset，则从 ibuf 数组的 0 位置开始填充。

DatagramPackage 提供的常用方法包括以下几种。

- public InetAddress getAddress()：如果是发送数据报，则获得数据报要发送的目标地址。如果是接收数据报，则返回发送此数据报的源地址。
- public byte[]getData()：返回一个字节数组，内容是数据报的数据。
- public int getLength()：获得数据报中数据的字节数。
- pubic int getPort()：返回数据报中的目标地址的主机端口号。

④ 使用 receive()方法接收发送方所发送的数据，同时这也是一个阻塞的方法。

⑤ 处理发送过来的数据。

下面的代码演示了处理接收数据的全部流程：

```
public void StartListen() {
    Integer port = 8903;
    byte[] message = new byte[100];

    try {
        DatagramSocket datagramSocket = new DatagramSocket(port);
        datagramSocket.setBroadcast(true);
        DatagramPacket datagramPacket = new DatagramPacket(message,
                message.length);
        try {
            while (!IsThreadDisable) {
                Log.d("UDP Demo", "准备接收");
                this.lock.acquire();

                datagramSocket.receive(datagramPacket);
                String strMsg = new String(datagramPacket.getData()).trim();
                Log.d("UDP Demo", datagramPacket.getAddress()
                        .getHostAddress().toString()
                        + ":" + strMsg );this.lock.release();
            }
        } catch (IOException e) {//IOException
            e.printStackTrace();
        }
    } catch (SocketException e) {
        e.printStackTrace();
    }

}
```

2. 发送方编程

发送方编程的一般步骤如下：

① 创建一个 DatagramSocket 对象。

② 创建一个 InetAddress，表示数据报的目的地址。

③ 准备发送数据，并将数据转换为 byte 类型。

④ 创建一个 DatagramPacket 对象，并指定数据报发送的目的地址及端口号。

⑤ 调用 DatagramSocket 对象的 send()方法发送数据。

下面的代码演示了处理发送数据的全部流程：

```
public void send(String message) {
    message = (message == null ? "Hello Android!" : message);
    int server_port = 8904;
    Log.d("UDP Demo", "UDP 发送数据:" + message);
    DatagramSocket s = null;

    try {
        s = new DatagramSocket();
    } catch (SocketException e) {
```

```
            e.printStackTrace();
        }

        InetAddress local = null;
        try {
            local = InetAddress.getByName("255.255.255.255");
        } catch (UnknownHostException e) {
            e.printStackTrace();
        }

        int msg_length = message.length();
        byte[] messageByte = message.getBytes();
        DatagramPacket p = new DatagramPacket(messageByte, msg_length, local,
                server_port);

        try {
            s.send(p);
            s.close();
        } catch (IOException e) {
            e.printStackTrace();
        }
}
```

注意,Google 不鼓励在手机中直接接收 UDP 包(手机开启 UDP 广播功能不仅耗电,而且占用系统资源),因此,有的手机厂商在定制 Rom 的时候关掉了这个功能。下面给出一个解决方案。

首先在 Activity 的 onCreate()方法里实例化一个 WifiManager. MulticastLock 对象 lock,例如:

```
WifiManager manager = (WifiManager)
        this.getSystemService(Context.WIFI_SERVICE);
WifiManager.MulticastLock lock = manager.createMulticastLock("wifi");
```

然后在调用广播发送、接收报文之前先调用 lock. acquire()方法,调用完之后及时调用 lock. release()方法释放资源(多次调用 lock. acquire()方法,可能会产生 java. lang. UnsupportedOperationException 异常)。

最后在配置文件里面添加如下权限:

```
<uses-permission
        android:name="android.permission.CHANGE_WIFI_MULTICAST_STATE" />
```

经过以上处理后,多数手机都能正常发送和接收广播报文。

4.4.2 NIO 编程

JDK 1.4 中引入的新输入输出库(New I/O,NIO,一种对 N 的理解是 Noblocking,即非阻塞的意思)在标准 Java 代码中提供了高速的、面向块的 I/O。

在使用 Socket 进行数据交换时,当 Socket 连接建立成功后,服务端和客户端都会拥有一个 Socket 实例,每个 Socket 实例都有一个 InputStream 和 OutputStream,正是通过这两个对象来交换数据。当 Socket 对象创建时,操作系统将会为 InputStream 和 OutputStream 分别分配一定大小的缓冲区,数据的写入和读取都是通过这个缓存区完成的。值得特别注意的是,这个缓存区的大小以及写入端的速度和读取端的速度非常影响这个连接的数据传输效率,如果两边同时传送数据时可能会产生死锁。

原来的 I/O 库(在 java.io.* 中)与 NIO 最重要的区别是数据打包和传输的方式。原来的 I/O 以流的方式处理数据,而 NIO 以块的方式处理数据。面向流的 I/O 系统一次一个字节地处理数据。一个输入流产生一个字节的数据,一个输出流消费一个字节的数据。面向块的 I/O 系统以块的形式处理数据。每一个操作都在一步中产生或者消费一个数据块。按块处理数据比按字节处理数据要快得多。但是面向块的 I/O 缺少一些面向流的 I/O 所具有的优雅性和简单性。

1. NIO 的核心类

(1) ByteBuffer

ByteBuffer 位于 java.nio 包中,目前提供了 Java 基本类型中除 Boolean 外其他类型的缓冲类型,如 ByteBuffer、DoubleBuffer、FloatBuffer、IntBuffer、LongBuffer 和 ShortBuffer 等。同时还提供了一种更特殊的映射字节缓冲类型 MappedByteBuffer。

使用 ByteBuffer 类的静态方法 static ByteBuffer allocate(int capacity)或 static ByteBuffer allocateDirect(int capacity)这两个方法来分配内存空间,两种方法的区别主要是后者更适用于反复分配的字节数组。

ByteBuffer 可以很好地和字节数组 byte[]进行转换类型,通过执行 ByteBuffer 类的 final byte[] array()方法就可以将 ByteBuffer 转为 byte[]。从 byte[]来构造 ByteBuffer 可以使用 wrap 方法,目前 Android 提供了两种重载方法,如 static ByteBuffer wrap(byte[] array)和 static ByteBuffer wrap(byte[] array, int start, int len)。第二个重载方法中第二个参数为 array 字节数组的起初位置,第三个参数为 array 字节数组的长度。

ByteBuffer 提供了多种 put/get 方法类型来添加/获取数据元素,如 put(byte b)、getDouble(int index)等。需要注意的是,按照 Java 的类型长度,一个 byte 占 1 字节,一个 char 类型是 2 字节,一个 float 或 int 是 4 字节,一个 long 则为 8 字节,所以内部的相关位置也会发生变化,同时每种方法还提供了定位的方法,如 ByteBuffer put(int index, byte b)。

(2) FileChannel

Channel 是一个对象,可以通过它读取和写入数据。在 NIO 中,除了 Socket 外,还提供了 File 设备的通道类,FileChannel 位于 java.nio.channels.FileChannel 包中。

通道与流的不同之处在于通道是双向的。而流只是在一个方向上移动(一个流必须是 InputStream 或者 OutputStream 的子类),而通道可以用于读、写或者同时用于读写。

下面的代码演示了 FileChannel 的使用。

```
String infile = "/sdcard/cwj.dat";
String outfile = "/sdcard/test.dat";

```

```
FileInputStream fin = new FileInputStream( infile );
FileOutputStream fout = new FileOutputStream( outfile );

FileChannel fcin = fin.getChannel();
FileChannel fcout = fout.getChannel();

ByteBuffer buffer = ByteBuffer.allocate( 1024 ); //分配 1KB 作为缓冲区

while (true) {
    buffer.clear(); //每次使用必须置空缓冲区

    int r = fcin.read( buffer );

    if (r == -1) {
        break;
    }

    buffer.flip();

    fcout.write( buffer );
}
```

flip()和 clear()这两个方法是 java.nio.Buffer 类中的方法,ByteBuffer 的父类是从 Buffer 类继承而来的。

(3) SocketChannel 与 ServerSocketChannel

在 Java 的 New I/O 中,处理 Socket 通信的是 SocketChannel,SocketChannel 关联了一个 Socket 类,使用 SocketChannel 类的 socket()方法可以返回一个传统 I/O 的 Socket 类。SocketChannel()对象在 Server 中一般通过 Socket 类的 getChannel()方法获得。

处理 ServerSocket 通信的是 ServerSocketChannel,通过 ServerSocketChannel 类的 socket()方法可以获得一个传统的 ServerSocket 对象,同时从 ServerSocket 对象的 getChannel()方法可以获得一个 ServerSocketChannel()对象,这点说明 NIO 的 ServerSocketChannel 和传统 IO 的 ServerSocket 是有关联的,实例化 ServerSocketChannel 只需要直接调用 ServerSocketChannel 类的静态方法 open()即可。

(4) Selector

New I/O 中的核心对象名为 Selector。Selector 就是注册对各种 I/O 事件感兴趣的地方,而且当那些事件发生时,就是由这个对象来反映所发生的事件。

调用 Selector 的静态工厂创建一个选择器,创建一个服务端的 Channel 并绑定到一个 Socket 对象,再把这个通信信道注册到选择器上,同时把这个通信信道设置为非阻塞模式。然后就可以调用 Selector 的 selectedKeys()方法来检查已经注册到这个选择器上的所有通信信道是否有需要的事件发生,如果有某个事件发生时,将会返回所有的 SelectionKey,通过这个对象 Channel 方法就可以取得这个通信信道对象,从而可以读取通信的数据,而这里读取的数据是 Buffer。例如:

```
public void selector() throws IOException {
    ByteBuffer buffer = ByteBuffer.allocate(1024);
    Selector selector = Selector.open();
    ServerSocketChannel ssc = ServerSocketChannel.open();
    ssc.configureBlocking(false);                          //设置为非阻塞方式
    ssc.socket().bind(new InetSocketAddress(8080));
    ssc.register(selector, SelectionKey.OP_ACCEPT);        //注册监听的事件
    while (true) {
        Set selectedKeys = selector.selectedKeys();        //取得所有 key 集合
        Iterator it = selectedKeys.iterator();
        while (it.hasNext()) {
            SelectionKey key = (SelectionKey) it.next();
            if ((key.readyOps() & SelectionKey.OP_ACCEPT) ==
                    SelectionKey.OP_ACCEPT) {
                ServerSocketChannel ssChannel =
                    (ServerSocketChannel) key.channel();
                SocketChannel sc = ssChannel.accept();  //接收到服务端的请求
                sc.configureBlocking(false);
                sc.register(selector, SelectionKey.OP_READ);
                it.remove();
            } else if ((key.readyOps() & SelectionKey.OP_READ) ==
                    SelectionKey.OP_READ) {
                SocketChannel sc = (SocketChannel) key.channel();
                while (true) {
                    buffer.clear();
                    int n = sc.read(buffer);                //读取数据
                    if (n <= 0) {
                        break;
                    }
                    buffer.flip();
                }
                it.remove();
            }
        }
    }
}
```

在上面的这段程序中，是将 Server 端的监听连接请求的事件和处理请求的事件放在一个线程中，但是在实际应用中，通常会把它们放在两个线程中，一个线程专门负责监听客户端的连接请求，而且是阻塞方式执行的；另外一个线程专门来处理请求，这个专门处理请求的线程才会真正采用 NIO 的方式。

2. 实现 NIO 编程

一般来说，很少在 Android 客户端实现 NIO 编程，毕竟 NIO 相对于 BIO 在逻辑处理上要复杂得多。下面介绍在 Android 平台上实现一个非阻塞的服务器的开发过程。

基本步骤如下：

① 通过 Selector 类的 open()静态方法实例化一个 Selector 对象。

② 通过 ServerSocketChannel 类的 open()静态方法实例化一个 ServerSocketChannel

对象。

③ 显式地调用 ServerSocketChannel 对象的 configureBlocking(false)方法,并设置为非阻塞模式。

④ 使用 ServerSocketChannel 对象的 socket()方法返回一个 ServerSocket 对象,使用 ServerSocket 对象的 bind()方法绑定一个 IP 地址和端口号。

⑤ 调用 ServerSocketChannel 对象的 register()方法注册相关的网络事件。

⑥ 通过 Selector 对象的 select()方法判断是否有关注的事件发生。

⑦ 如果 Selector 对象的 select()方法返回的结果数大于 0,则通过 Selector 对象的 selectedKeys()方法获取一个 SelectionKey 类型的 Set 集合,使用 Java 的迭代器 Iterator 类来遍历这个 Set 集合(注意判断 SelectionKey 对象)。

⑧ 处理完 SelectionKey 对象后需要从 Set 集合中移除。

⑨ 接下来判断 SelectionKey 对象的事件,使用 SelectionKey 对象的 isAcceptable()方法进行判断。

下面的 connect()方法演示了上述过程。

```
public void connect() {
    Selector selector = null;
    ServerSocketChannel ssc = null;

    try {
        selector = Selector.open();
        ssc = ServerSocketChannel.open();
        ssc.socket().bind(new InetSocketAddress(1988));
        ssc.configureBlocking(false);
        ssc.register(selector, SelectionKey.OP_ACCEPT);

        while (true) {
            int n = selector.select();
            if (n < 1)
                continue;
            Iterator<SelectionKey> it = selector.selectedKeys().iterator();
            while (it.hasNext()) {
                SelectionKey key = it.next();
                it.remove();

                if (key.isAcceptable()) {
                    ServerSocketChannel ssc2 = (ServerSocketChannel) key
                            .channel();
                    SocketChannel channel = ssc2.accept();
                    channel.configureBlocking(false);
                    channel.register(selector, SelectionKey.OP_READ);

                    Log.i("CWJ Client :", channel.socket().getInetAddress()
                            .getHostName() + ":" + channel.socket().getPort());
                }

```

```
                else if (key.isReadable()) {
                    SocketChannel channel = (SocketChannel) key.channel();
                    ByteBuffer buffer = ByteBuffer.allocate(1024);
                    channel.read(buffer);
                    buffer.flip();
                    Log.i("receive info:", buffer.toString());
                    channel.write(CharBuffer.wrap("it works".getBytes()));
                }
            }
        }
    } catch (IOException e) {
        e.printStackTrace();
    } finally {
        try {
            selector.close();
            server.close();
        } catch (IOException e) {
        }
    }
}
```

上面的方法在单次交互以及数据量较小时一般没有太大问题，但是对于多次交互及数据量大时，也可能发生阻塞。例如，在判断 key.isReadable()时，对于这个 SelectionKey 关联的 SocketChannel，尽量不要在写入数据量过多时，让 ByteBuffer 调用 hasRemaining()这样的方法。读者可以仔细阅读资源包中的 SocketUtil 以了解健壮性更强的 NIO 方法。

4.5 习　　题

1. 完善 4.3 节 Socket 编程的示例，实现 Socket 与 ServerSocket 之间的相互通信。
2. 使用 UPD 编程实现手机和计算机之间的相互通信。

第 5 章　HTTP 编程

大多数网络连接的 Android 应用使用 HTTP 发送和接收数据，Android 包括两个 HTTP 客户端：HttpURLConnection 和 Apache HttpClient，它们支持 HTTPS、流的上传和下载，可配置 IPv6 以及连接池。

5.1　HTTP 协议与 URL

本节介绍关于 HTTP 协议的基础知识，以及 URL 的组成。

5.1.1　HTTP 协议

HTTP(HyperText Transfer Protocol，超文本传输协议)是一个基于请求与响应模式的、无状态的应用层协议，通常基于 TCP 的连接方式。

1. HTTP 协议的特点与分类

HTTP 协议的主要特点如下。

- 支持客户/服务器模式。
- 简单快速：客户向服务器请求服务时，只需传送请求方法和路径。
- 灵活：HTTP 允许传输任意类型的数据对象(类型由 Content-Type 加以标记)。
- 无连接：即每次连接只处理一个请求，处理完客户的请求，并收到客户的应答后，即断开连接。采用这种方式可以节省传输时间。
- 无状态：无状态是指协议对于事务处理没有记忆能力。

HTTP 协议包括以下两种。

- HTTP 1.0 协议：客户端在每次向服务器发出请求后，服务器就会向客户端返回响应消息(包括请求是否正确以及所请求的数据)，在确认客户端已经收到响应消息后，服务端就会关闭网络连接(即关闭 TCP 连接)。在这个数据传输过程中，并不保存任何历史信息和状态信息，因此，HTTP 1.0 协议也被认为是无状态的协议。
- HTTP 1.1 协议：当客户端连接到服务器后，服务器就将关闭客户端连接的主动权交还给客户端。也就是说，在客户端向服务器发送一个请求并接收一个响应后，只要不调用类似 Socket 类的 close()方法关闭网络连接，就可以继续向服务器发送 HTTP 请求。并且同一对客户/服务器之间的后续请求和响应都可以通过这个连接发送，这样就可以大大减轻服务器的压力。

2. HTTP 请求/响应的组成

HTTP 消息由客户端到服务器的请求和服务器到客户端的响应组成。请求消息和响应消息都是由开始行(对于请求消息,开始行就是请求行;对于响应消息,开始行就是状态行)、消息报头(可选)、空行(只有 CRLF 的行)、消息正文(可选)组成。

(1) 请求行

请求行以一个方法符号开头,以空格分开,后面跟着请求的 URI 和协议的版本,格式如下:

```
Method Request - URI HTTP - Version CRLF
```

其中 Method 表示请求方法;Request-URI 是一个统一资源标识符;HTTP-Version 表示请求的 HTTP 协议版本;CRLF 表示回车和换行(除了作为结尾的 CRLF 外,不允许出现单独的 CR 或 LF 字符)。

请求方法(所有方法全为大写)解释如表 5-1 所示。

表 5-1 HTTP 的请求方法

请求方法	说明
GET	请求获取 Request-URI 所标识的资源
POST	在 Request-URI 所标识的资源后附加新的数据
HEAD	请求获取由 Request-URI 所标识的资源的响应消息报头
PUT	请求服务器存储一个资源,并用 Request-URI 作为其标识
DELETE	请求服务器删除 Request-URI 所标识的资源
TRACE	请求服务器回送收到的请求信息,主要用于测试或诊断
CONNECT	保留将来使用
OPTIONS	请求查询服务器的性能,或者查询与资源相关的选项和需求

(2) 状态行

状态行包括 HTTP 协议版本号、状态码、状态码的文本描述信息。如:HTTP/1.1 200 OK。

状态码由一个三位数组成,状态码大体有 5 种含义。

- 1xx:信息,请求收到,继续处理。
- 2xx:成功。200 表示请求成功;206 表示断点续传。
- 3xx:重定向。一般跳转到新的地址。
- 4xx:客户端错误。404 表示文件不存在。
- 5xx:服务器错误。500 表示内部错误。

(3) 消息报头

HTTP 消息报头包括普通报头、请求报头、响应报头、实体报头。每一个报头域都是由“名字+':'+空格+值”组成的,消息报头域的名字与大小写无关。

① 请求报头

请求报头允许客户端向服务器端传递请求的附加信息以及客户端自身的信息。

常用的请求报头包括以下几种。

- Accept:请求报头域用于指定客户端接收哪些类型的信息。
- Accept-Charset:请求报头域用于指定客户端接收的字符集。
- Accept-Encoding:请求报头域类似于 Accept,但是它是用于指定可接收的内容编码。

- Accept-Language：请求报头域类似于 Accept，但是它是用于指定一种自然语言。
- Authorization：请求报头域主要用于证明客户端有权查看某个资源。
- Host：请求报头域主要用于指定被请求资源的 Internet 主机和端口号，它通常从 HTTP URL 中提取出来。
- User-Agent：请求报头域允许客户端将它的操作系统、浏览器和其他属性告诉服务器。

② 响应报头

响应报头允许服务器传递不能放在状态行中的附加响应信息，以及关于服务器的信息和对 Request-URI 所标识的资源进行下一步访问的信息。

常用的响应报头包括以下几种。

- Location：响应报头域用于重定向接收者到一个新的位置。Location 响应报头域常用在更换域名的时候。
- Server：响应报头域包含了服务器用来处理请求的软件信息。

③ 实体报头

请求和响应消息都可以传送一个实体。

常用的实体报头包括以下几种。

- Content-Encoding：指示已经被应用到实体正文的附加内容的编码。
- Content-Language：实体报头域描述了资源所用的自然语言。
- Content-Length：实体报头域用于指明实体正文的长度，以字节方式存储的十进制数字来表示。
- Content-Type：实体报头域用于指明发送给接收者的实体正文的媒体类型。
- Last-Modified：实体报头域用于指示资源的最后修改日期和时间。
- Expires：实体报头域给出响应过期的日期和时间。

5.1.2 URL

URL(Uniform Resource Locator)代表一个统一资源定位符，它是指向互联网“资源”的指针。资源可以是简单的文件或目录，也可以是对更为复杂的对象的引用，例如对数据库或搜索引擎的查询。

1. URL 的组成

URL 一般包含协议、主机名、端口、路径、查询字符串和参数等对象。URL 的格式如下：

```
protocol://host:[port]/path/[parameters][?query]
```

- protocol(协议)：最常用的是 HTTP 协议，它也是目前 WWW 中应用最广的协议，格式为“http://”。ftp 通过 FTP 访问资源，格式为“ftp://”。mailto 资源为电子邮件地址，通过 SMTP 访问，格式为“mailto://”。
- hostname(主机名)：是指存放资源的服务器的域名系统(DNS)主机名或 IP 地址。
- port(端口号)：整数，可选，省略时使用方案的默认端口，各种传输协议都有默认的端口号，如 http 的默认端口为 80。

- path(路径)：由零或多个“/”符号隔开的字符串，一般用来表示主机上的一个目录或文件地址。
- parameters：资源名称等参数。
- ? query(查询)：用于给动态网页传递参数，可有多个参数，用“&”符号隔开，每个参数的名和值用“=”符号隔开。

例如，http://www.oracle.com/index.html 表示该 URL 使用的协议为 http(超文本传输协议)，并且该信息驻留在一台名为 www.oracle.com 的主机上。主机上的信息名称为“/index.html”，这一部分称为路径部分。

应用程序也可以指定一个“相对 URL”，它只包含到达相对于另一个 URL 的资源的足够信息。HTML 页面中经常使用相对 URL。相对 URL 不需要指定 URL 的所有组成部分。如果缺少协议、主机名称或端口号，这些值将从完整指定的 URL 中继承。例如，faq.html 即为 http://www.oracle.com/faq.html 的缩写。

下面的代码演示了 Java 所支持的 URL 类型：

```
public void supportProtocal() {

    String host = "www.google.com";
    String file = "/index.html";

    String[] schemes = {"http", "https", "ftp", "mailto", "telnet", "file",
            "ldap", "gopher", "jdbc", "rmi", "jndi", "jar", "doc", "netdoc",
            "nfs", "verbatim", "finger", "daytime", "systemresource"};

    for (int i = 0; i < schemes.length; i++) {
        try {
            URL u = new URL(schemes, host, file);
            Log.i("supportProtocal", schemes + " is supported/r/n");
        } catch (Exception ex) {
            Log.i("supportProtocal", schemes + " is not supported/r/n");
        }
    }
}
```

下面介绍 URL 的编码与解码。

当 URL 地址里包含非西欧字符的字符串时，URL 并不自动执行编码或解码工作。java.net 包中的 URLEncoder 和 URLDecoder 类提供了编码与解码的方法，如下所示。

- encode(String s, String enc)：使用指定的编码机制将字符串转换为 application/x-www-form-urlencoded 格式。
- decode(String s, String enc)：使用指定的编码机制对 application/x-www-form-urlencoded 字符串解码。

2. 主要方法

URL 包含的主要方法包括以下几种。

- URL(String spec)：根据 String 表示形式创建 URL 对象。生成 URL 对象时，必须要用 try-catch 语句进行 MalformedURLException 例外捕获。
- URL(String protocol, String host, int port, String file)：根据指定 protocol、host、port 和 file 创建 URL 对象。URL 提供了 getProtocol()、getHost()、getPort()等方法来获取 URL 的组成部分。
- openConnection()：返回一个 URLConnection 对象，它表示到 URL 所引用的远程对象的连接。每次调用 openConnection()方法都打开一个新的连接。如果 URL 的协议（例如 HTTP 或 JAR）存在属于以下包或其子包之一的公共、专用 URLConnection 子类(包括 java.lang、java.io、java.util、java.net)，返回的连接将为该子类的类型。例如，对于 HTTP，将返回 HttpURLConnection；对于 JAR，将返回 JarURLConnection。
- openConnection(Proxy proxy)：与 openConnection()类似，所不同的是连接通过指定的代理建立；不支持代理方式的协议处理程序将忽略该代理参数并建立正常的连接。
- openStream()：打开到此 URL 的连接并返回一个用于从该连接读入的 InputStream。
- setURLStreamHandlerFactory(URLStreamHandlerFactory fac)：设置应用程序的 URLStreamHandlerFactory。在一个给定的 Java 虚拟机中，此方法最多只能调用一次。URLStreamHandlerFactory 实例用于从协议名称构造流协议处理程序。如果有安全管理器，此方法首先调用安全管理器的 checkSetFactory 方法以确保允许该操作。这可能会导致 SecurityException 异常。
- toString()：构造此 URL 的字符串表示形式。
- toURI()：返回与此 URL 等效的 URI。此方法的作用与 new URI (this.toString())相同。注意，任何 URL 实例只要遵守 RFC 2396 就可以转化为 URI。但是，有些未严格遵守该规则的 URL 将无法转化为 URI。

使用 java.io 流处理类从 URL 中读取数据是一个非常简单的过程。一旦建立了一个成功的连接，就可以获得针对这个连接的输入流并且开始进行读操作。下面的示例展示了如何从 URL 中读取文本数据。

```
public void readTxt(String _url) {
    try {
        URL url = new URL(_url);
        InputStream is = url.openStream();
        InputStreamReader isr = new InputStreamReader(is);
        BufferedReader bf = new BufferedReader(isr);
        String str;

        while ((str = bf.readLine()) != null) {
            Log.i(TAG, str);
        }
    } catch (MalformedURLException e) {
        e.printStackTrace();
    } catch (IOException e) {
```

```
            e.printStackTrace();
        }
}
```

5.2 HttpURLConnection 编程

java.net.HttpURLConnection 类是继承自 URLConnection(抽象类 URLConnection 是所有类的超类,它代表应用程序和 URL 之间的通信连接。此类的实例可用于读取和写入此 URL 引用的资源。)的一个抽象类,是一种多用途、轻量极的 HTTP 客户端,使用它来进行 HTTP 操作可以适用于大多数的应用程序。

5.2.1 创建 HttpURLConnection 连接

通常创建一个和 URL 的连接,并发送请求、读取此 URL 引用的资源需要如下几个步骤:

① 通过 URL 对象的 openConnection()方法来创建 URLConnection 对象。

② 设置 URLConnection 的参数和普通请求属性。

③ 如果是发送 GET 方式的请求,使用 connect()方法建立和远程资源之间的实际连接即可;如果需要发送 POST 方式的请求,需要获取 URLConnection 实例对应的输出流来发送请求参数。

④ 远程资源变为可用,程序可以访问远程资源的头字段,或通过输入流来读取远程资源的数据。

1. 创建 HttpURLConnection 对象

HttpURLConnection 是一种访问 HTTP 资源的方式,在 HTTP 编程时,来自 HttpURLConnection 的类是所有操作的基础。

HttpURLConnection 是一个抽象类,不能通过 new HttpURLConnection()的方式来获取一个 HttpURLConnection 对象。常见的做法是使用 java.net.URL 封装 HTTP 资源的 URL,并使用 openConnection()方法获得 HttpURLConnection 对象,例如:

```
try {
    URL url = new URL(httpUrl);
    URLConnection urlConnection = url.openConnection();
    HttpURLConnection httpUrlConnection = (HttpURLConnection) urlConnection;
    } catch (MalformedURLException e) {
    e.printStackTrace();
    } catch (IOException e) {
    e.printStackTrace();
}
```

2. 设置 HttpURLConnection 参数

HttpURLConnection API 提供了一系列的 set 方法来设置网络连接的参数,主要有以下几种。

- void setConnectTimeout(int timeout)：设置一个指定的超时值(以毫秒为单位),该值将在打开到此 URLConnection 引用的资源的通信连接时使用。如果在建立连接之前超时期满,则会引发一个 java.net.SocketTimeoutException。超时时间为 0 表示无穷大超时。
- void setRequestMethod(String method)：设置 URL 请求的方法,包括 GET、POST、HEAD、OPTIONS、PUT、DELETE 和 TRACE,具体取决于协议的限制。默认方法为 GET。如果无法重置方法或者请求的方法对 HTTP 无效,则抛出 ProtocolException 异常。
- void setDoInput(boolean doinput)：将此 URLConnection 的 doInput 字段的值设置为指定的值。URL 连接可用于输入或输出。如果打算使用 URL 连接进行输入,则将 DoInput 标志设置为 true；如果不打算使用,则设置为 false。默认值为 true。
- void setDoOutput(boolean dooutput)：将此 URLConnection 的 doOutput 字段的值设置为指定的值。如果打算使用 URL 连接进行输出,则将 DoOutput 标志设置为 true；如果不打算使用,则设置为 false。默认值为 false。
- void setDefaultUseCaches(boolean defaultusecaches)：将 useCaches 字段的默认值设置为指定的值。
- void setUseCaches(boolean usecaches)：将此 URLConnection 的 useCaches 字段的值设置为指定的值。有些协议用于文档缓存。有时候能够进行"直通"并忽略缓存尤其重要,例如浏览器中的"重新加载"按钮。如果连接中的 UseCaches 标志为 true,则允许连接使用任何可用的缓存；如果为 false,则忽略缓存；默认值来自 DefaultUseCaches,默认为 true。
- void setRequestProperty(String key, String value)：设置一般请求属性。key 用于识别请求的关键字(例如"Content-type")。value 表示与该键关联的值(例如"application/x-java-serialized-object")。HTTP 要求所有能够合法拥有多个具有相同键的实例的请求属性,使用以逗号分隔的列表语法,这样可将多个属性添加到一个属性中。
- static void setContentHandlerFactory(ContentHandlerFactory fac)：设置应用程序的 ContentHandlerFactory。一个应用程序最多只能调用一次该方法。ContentHandlerFactory 实例用于根据内容类型构造内容处理程序。如果有安全管理器,此方法首先调用安全管理器的 checkSetFactory 方法以确保允许该操作。这可能会导致 SecurityException 异常。
- static void setDefaultAllowUserInteraction(boolean defaultallowuserinteraction)：将未来所有 URLConnection 对象的 allowUserInteraction 字段默认值设置为指定的值。
- void setChunkedStreamingMode(int chunklen)：此方法用于在预先不知道内容长度时启用没有进行内部缓冲的 HTTP 请求正文的流。在此模式下,使用存储块传输编码发送请求正文。注意,并非所有 HTTP 服务器都支持此模式。启用输出流时,不能自动处理验证和重定向。如果需要验证和重定向,则在读取响应时将抛出

HttpRetryException。可以查询此异常以获取错误的详细信息。该方法必须在连接 URLConnection 前调用。

- void setFixedLengthStreamingMode(int contentLength)：此方法用于在预先已知内容长度时启用没有进行内部缓冲的 HTTP 请求正文的流。如果应用程序尝试写入的数据多于指示的内容长度，或者应用程序在写入指示的内容长度前关闭了 OutputStream，将抛出异常。启用输出流时，不能自动处理验证和重定向。如果需要验证和重定向，则在读取响应时将抛出 HttpRetryException。可以查询此异常以获取错误的详细信息。该方法必须在连接 URLConnection 前调用。

3. HttpURLConnection 连接

当设置好 HttpURLConnection 的连接参数后，即可以通过 connect()方法进行网络连接。如果在已打开连接(此时 connected()方法返回值为 true)的情况下调用 connect()方法，则忽略该调用。

注意：HttpURLConnection 的 connect()方法实际上只是建立了一个与服务器的 tcp 连接，并没有实际发送 http 请求。无论是 post 请求方式还是 get 请求方式，http 请求实际上直到 HttpURLConnection 的 getInputStream()这个方法里面才正式发送出去。

如果服务器近期不太可能有其他请求，可以调用 disconnect()断开连接。disconnect()并不意味着可以对其他请求重用此 HttpURLConnection 实例。

在使用 HttpURLConnection 进行网络连接时需要注意：

- 一般需要通过 setConnectTimeout()方法设置连接超时，如果网络不好，Android 系统在超过默认时间后会收回资源，中断操作。
- 通过 getResponseCode()对响应码进行判断。如果返回的响应码是 200，则表示连接成功。
- 在对大文件进行操作时，要将文件写到 SDCard 上，不要直接写到手机内存上。
- 操作大文件时，要一边从网络上读取，一边往 SDCard 上写，以便减少手机内存的使用。
- 对文件流操作完毕后要记得及时关闭。

5.2.2 HttpURLConnection 数据交换

HttpURLConnection API 提供了一系列的 get 方法来获取网络传递的信息，主要有以下几种。

- Object getContent()：获取该 URLConnection 的内容。
- String getHeaderField(String name)：获取指定响应头字段的值。经常有可能访问的头字段有如下内容。
  - ➢ getContentEncoding()：获取 content-enconding 响应头字段的值。
  - ➢ getContentLength()：获取 content-length 响应头字段的值。
  - ➢ getContentType()：获取 contet-type 响应头字段的值。
  - ➢ getDate()：获取 date 响应头字段的值。
  - ➢ getExpiration()：获取 expires 响应头字段的值。
  - ➢ getLastModified()：获取 last-modified 响应头字段的值。

- InputStream getInputStream()：返回从此打开的连接读取的输入流。在读取返回的输入流时，如果在数据可供读取之前达到读入超时时间，则会抛出 SocketTimeoutException。
- OutputStream getOutputStream()：返回写入此连接的输出流。
- int getResponseCode()：从 HTTP 响应消息获取状态码。如果无法从响应中识别任何代码(即响应不是有效的 HTTP)，则返回－1。常见的状态码包括 HTTP_OK(状态码为 200，表示服务器成功返回网页)、HTTP_NOT_FOUND(状态码为 404，表示请求的网页不存在)、HTTP_SERVICE_UNAVAILABLE(状态码为 503，表示服务器超时)等。
- String getResponseMessage()：获取与来自服务器的响应代码一起返回的 HTTP 响应消息(如果有)。如果无法从响应识别任何字符(结果不是有效的 HTTP)，则返回 null。
- InputStream getErrorStream()：如果连接失败但服务器仍然发送了有用数据，则返回错误流。典型示例是，当 HTTP 服务器使用 404 响应时，将导致在连接中抛出 FileNotFoundException，但是服务器同时还会发送建议如何操作的 HTML 帮助页。此方法不会导致启用连接。如果没有建立连接，或者在连接时服务器没有发生错误，或服务器出错但没有发送错误数据，则此方法返回 null。这是默认设置。

对 HTTP 资源的读写操作是通过 InputStream 和 OutputStream 进行的，下面列举几个常见的用法。

1. 使用 POST 方式请求数据

使用 POST 方式请求数据的一般步骤如下。

(1) 确定 URL，一般结构为 uri。例如：

```
String uri = "http://localhost/njcit/login.jsp";
URL url = new URL(uri);
```

(2) 确定请求参数。

```
String params = "userName = value1&loginPwd = value2& … ";
```

(3) 通过 URL 创建 HttpURLConnection 对象。

(4) HttpURLConnection 设置连接可读写数据。例如：

```
conn.setDoOutput(true);
conn.setDoInput(true);
```

(5) 通过 getOutputStream()获得输出流对象，进而发送请求参数。例如：

```
Printer out = new Printer(conn.getOutputStream());
out.write(params);
```

下面的代码演示了使用 POST 方式发送数据的过程：

```
public static String doPost(String url, String param) {
    PrintWriter out = null;
    BufferedReader in = null;
    String result = "";
    try {
        URL realUrl = new URL(url);
        //打开和 URL 之间的连接
        HttpURLConnection conn = (HttpURLConnection) realUrl
                .openConnection();
        //设置通用的请求属性
        conn.setRequestProperty("accept", "*/*");
        conn.setRequestProperty("connection", "Keep-Alive");
        conn.setRequestMethod("POST");
        conn.setRequestProperty("Content-Type",
                "application/x-www-form-urlencoded");
        conn.setRequestProperty("charset", "utf-8");
        conn.setUseCaches(false);
        //发送 POST 请求必须设置如下两行
        conn.setDoOutput(true);
        conn.setDoInput(true);
        conn.setReadTimeout(TIMEOUT_IN_MILLIONS);
        conn.setConnectTimeout(TIMEOUT_IN_MILLIONS);

        if (param != null && !param.trim().equals("")) {
            //获取 URLConnection 对象对应的输出流
            out = new PrintWriter(conn.getOutputStream());
            //发送请求参数
            out.print(param);
            //flush 输出流的缓冲
            out.flush();
        }
        //定义 BufferedReader 输入流来读取 URL 的响应
        in = new BufferedReader(
                new InputStreamReader(conn.getInputStream()));
        String line;
        while ((line = in.readLine()) != null) {
            result += line;
        }
    } catch (Exception e) {
        e.printStackTrace();
    }
    //使用 finally 块来关闭输出流、输入流
    finally {
        try {
            if (out != null) {
                out.close();
            }
            if (in != null) {
```

```
49                  in.close();
50              }
51          } catch (IOException ex) {
52              ex.printStackTrace();
53          }
54      }
55      return result;
56  }
```

代码分析：param 是请求参数，请求参数的一般形式是 name1＝value1&name2＝value2。14～15 行设置本次连接的 Content-type 为 application/x-www-form-urlencoded，表示正文是 UrlEncoded 编码过的 form 参数。17 行设置 POST 请求不能使用缓存。19 行设置是否向 HttpURLConnection 输出，因为是 POST 请求，参数要放在 Http 正文内，因此需要设为 true。

HTTP 通信中使用最多的是 GET 和 POST 两种请求。GET 请求可以获取静态页面，也可以把参数放在 URL 字符串后面传递给服务器。POST 与 GET 的不同之处在于 POST 的参数不是放在 URL 字符串里面，而是放在 HTTP 请求数据中。使用 POST 请求时，不需要在 URL 中附加任何参数，这些参数会通过 cookie 或者 session 等其他方式以键值对的形式传送到服务器上。

注意：在用 POST 方式发送 URL 请求时，URL 请求参数的设定都必须要在 connect()方法执行之前完成。而对 OutputStream 的写操作，又必须要在 InputStream 的读操作之前。如果 InputStream 读操作在 OutputStream 的写操作之前，会抛出 java.net.ProtocolException 异常。

2. 使用 GET 方式请求数据

使用 GET 方式请求数据的一般步骤如下。

① 确定 URL，一般结构为 uri＋"?"＋params。例如：

```
String uri = "http://localhost/njcit/login.jsp";
String params = "userName = value1&loginPwd = value2& …";
URL url = new URL(uri + params);
```

② 通过 URL 调用 openConnection()方法创建 HttpURLConnection 对象。

③ HttpURLConnection 对象调用 setConnectTimeout(int milisecond)设置网络响应时间。调用 setRequestMethod(String method)设置发送请求的方法。

④ 通过 getInputStream()获得输入流，进而获得响应。

下面的代码演示了使用 GET 方式请求数据的过程：

```
public static String doGet(String urlStr) {
    URL url = null;
    HttpURLConnection conn = null;
    InputStream is = null;
```

```
    ByteArrayOutputStream baos = null;
    try {
        url = new URL(urlStr);
        conn = (HttpURLConnection) url.openConnection();
        conn.setReadTimeout(TIMEOUT_IN_MILLIONS);
        conn.setConnectTimeout(TIMEOUT_IN_MILLIONS);
        conn.setRequestMethod("GET");
        conn.setRequestProperty("accept", "*/*");
        conn.setRequestProperty("connection", "Keep-Alive");
        if (conn.getResponseCode() == 200) {
            is = conn.getInputStream();
            baos = new ByteArrayOutputStream();
            int len = -1;
            byte[] buf = new byte[128];

            while ((len = is.read(buf)) != -1) {
                baos.write(buf, 0, len);
            }
            baos.flush();
            return baos.toString();
        } else {
            throw new RuntimeException(" responseCode is not 200 ... ");
        }

    } catch (Exception e) {
        e.printStackTrace();
    } finally {
        try {
            if (is != null)
                is.close();
        } catch (IOException e) {
        }
        try {
            if (baos != null)
                baos.close();
        } catch (IOException e) {
        }
        conn.disconnect();
    }

    return null;
}
```

在 GET 请求时，一般在 URL 中带有请求的参数，请求的 URL 格式通常为："http://×××.××××.com/××.aspx? param=value")

在使用 HttpURLConnection 进行数据交换时，需要注意如下事项：

- 上传数据至服务器时（即向服务器发送请求），如果知道上传数据的大小，应该显式使用 setFixedLengthStreamingMode(int)来设置上传数据的精确值;如果不知道上传数据的大小,则应使用 setChunkedStreamingMode(int)(通常使用默认值 0 作为实际参数传入)。如果两个方法都未设置,则系统会强制将"请求体"中的所有内容都缓存至内存中(在通过网络进行传输之前),这样会浪费"堆"内存(甚至可能耗尽),并加重隐患。

- 如果通过流输入或输出少量数据,则需要使用带缓冲区的流(如 BufferedInputStream);大量读取或输出数据时,可忽略缓冲流(不使用缓冲流会增加磁盘 I/O,默认的流操作是直接进行磁盘 I/O 操作的)。
- 当需要传输(输入或输出)大量数据时,使用"流"来限制内存中的数据量,即将数据直接放在"流"中,而不是存储在字节数组或字符串中(这些都存储在内存中)。

5.3 HttpClient 编程

在 Android 开发中,Android SDK 附带了 Apache 的 Http 服务工具 HttpClient。Apache HttpClient 是一个开源项目,弥补了 java.net 灵活性不足的缺点,提供了对 HTTP 协议的全面支持,为客户端的 HTTP 编程提供高效、最新、功能丰富的工具包支持。

5.3.1 HttpClient 简介

HttpClient 是 Apache Jakarta Common 下的子项目,可以用来提供高效的、最新的、功能丰富的支持 HTTP 协议的客户端编程工具包,并且它支持 HTTP 协议最新的版本和建议。

1. 功能介绍

org.apache.http.client.HttpClient 接口提供的主要功能包括:

- 实现了所有 HTTP 的方法(GET、POST、PUT、HEAD 等)。
- 支持自动转向①。
- 支持 HTTPS 协议。
- 支持验证会话机制②。
- 支持代理服务器等。

2. 与 HttpURLConnection 对比

HttpURLConnection 和 HttpClient 都支持 HTTPS 协议、IPv6、以流的形式进行上传和下载、配置超时时间以及连接池等功能。两者的比较见表 5-2。

① HttpClient 支持自动转向处理,但是像 POST 和 PUT 这种要求接受后继服务的请求方式,暂时不支持自动转向,因此如果碰到 POST 方式提交后返回的是 301 或者 302,则需要自己处理。

② HttpClient 的好处在于,同一个手机客户端在登录验证通过后,可以保持这个客户的会话状态,在同一个站点的其他业务模块中进行新的业务实现时,可以视作同一个登录客户。该方式类似于 Web 应用中的 Session 会话机制,可以从反复的验证业务中解放出来,提高用户的体验。

表 5-2　HttpURLConnection 和 HttpClient 的比较

比较	HttpURLConnection	HttpClient
功能用法	• HttpURLConnection 对大部分工作进行了包装，屏蔽了不需要的细节，适合开发人员直接调用。 • HttpURLConnection 在 2.3 版本中增加了一些 HTTPS 方面的改进，4.0 版本增加了一些响应的缓存	• HttpClient 库要丰富很多，提供了很多工具，封装了 http 的请求头、参数、内容体、响应，还有一些高级功能，以及对代理、COOKIE、鉴权、压缩、连接池的处理。 • HttpClient 高级功能代码写起来比较复杂，对开发人员的要求会高一些
性能对比	• HttpUrlConnection 直接支持 GZIP 压缩。 • HttpUrlConnection 直接支持系统级连接池，即打开的连接不会直接关闭，在一段时间内所有程序可共用。 • HttpUrlConnection 直接在系统层面做了缓存策略处理（4.0 版本以上），加快了重复请求的速度	• HttpClient 也支持 GZIP 压缩，但要自己写代码处理。 • HttpClient 也能支持系统级连接池，但不如 HttpUrlConnection 直接进行系统的底层支持那样好
未来发展	• HttpURLConnect 是一个通用的、适合大多数应用的轻量级组件。这个类起步比较晚，很容易在主要 API 上做稳步的改善。但是 HttpURLConnection 在 Android 2.2 及以下版本上存在一些令人厌烦的 bug，尤其是在读取 InputStream 时调用 close()方法，就有可能会导致连接池失效了。 • Android 团队未来的工作会将更多的时间放在优化 HttpURLConnection 上，它的 API 简单，体积较小，因而非常适用于 Android 项目。压缩和缓存机制可以有效地减少网络访问的流量，在提升速度和省电方面也起到了较大的作用	HttpClient 适用于 Web 浏览器，它们是可扩展的，并且拥有大量稳定的 API。但是，在不破坏其兼容性的前提下很难对如此多的 API 做修改。因此，Android 团队对修改优化 Apache HTTP Client 表现得并不积极
选用建议	Android 2.3 及以上版本建议选用 HttpURLConnection，2.2 及以下版本建议选用 HttpClient。新的应用都建议使用 HttpURLConnection	如果一个 Android 应用需要向指定页面发送请求，但该页面并不是一个简单的页面，只有当用户已经登录，而且登录用户的用户名有效时才可访问该页面。如果使用 HttpURLConnection 来访问这个被保护的页面，那么需要处理的细节就太复杂了。这种情况建议使用 HttpClient

Google 在其 SDK 文档中通过博客说明，对于 Gingerbread 及其以后的版本来说，HttpURLConnection 是最好的选择，其简洁的 API 和轻量级的实现用于 Android 系统再适合不过了，同时，对开发者透明的压缩和缓存实现，可以减少网络数据传输量，提高程序响应速度，也节约设备电源。不过，HttpClient 封装了很多有用的工具，便于开发者使用，并提高开发效率。

3. 一般使用步骤

使用 HttpClient 实现 Http 编程的一般步骤如下：

① 使用 DefaultHttpClient 类实例化 HttpClient 对象。

② 如果需要发送 GET 请求,则创建 HttpGet 对象;如果需要发送 POST 请求,则创建 HttpPost 对象。

③ 如果需要发送请求参数,可以调用 HttpGet 以及 HttpPost 共同的 setParams (HttpParams params)方法来添加请求参数,HttpPost 要通过 setEntity(HttpEntity entity)的方式设置请求参数。

④ 将要请求的 URL 通过构造方法传入 HttpGet 或 HttpPost 对象。

⑤ 调用 HttpClient 的 execute()方法发送 HTTP GET 或 HTTP POST 请求,并返回一个响应对象 HttpResponse。

⑥ 通过 HttpResponse 接口的 getEntity()方法获得 HttpEntity 响应对象,该对象包装了响应内容,通过该对象的解析可以得到响应内容。

⑦ 通过响应对象 HttpResponse 的 getAllHeaders()、getHeaders(String name)等方法获得服务响应头。

下面的 AsyncTask 实现类代码演示了使用 HttpPost 登录人人网的功能。

```
class PostLoader extends AsyncTask<Void, Integer, String> {

    @Override
    protected String doInBackground(Void... params) {
        String httpUrl = "http://3g.renren.com/login.do";
        String strResult = "";

        HttpPost httpRequest = new HttpPost(httpUrl);
        List<NameValuePair> _params = new ArrayList<NameValuePair>();
        _params.add(new BasicNameValuePair("email", "******"));
        _params.add(new BasicNameValuePair("&password", "******"));
        try {
            HttpEntity httpentity = new UrlEncodedFormEntity(_params,
                    "UTF-8");
            httpRequest.setEntity(httpentity);

            HttpClient httpclient = new DefaultHttpClient();
            HttpResponse httpResponse = httpclient.execute(httpRequest);
            if (httpResponse.getStatusLine().getStatusCode() ==
                    HttpStatus.SC_OK) {
                strResult = EntityUtils.toString(httpResponse.getEntity());
            }
        } catch (UnsupportedEncodingException e) {
        } catch (ClientProtocolException e) {
        } catch (ParseException e) {
        } catch (IOException e) {
        }

        return strResult;
    }

    @Override
    protected void onPostExecute(String result) {
```

```
        mWebView.loadData(result, "text/html; charset = UTF - 8", null);
    }
}
```

4. 单实例模式

在实际项目中，可能在多处需要进行 HTTP 通信，这时候不需要为每个请求都创建一个新的 HttpClient。对于一个通信单元甚至是整个应用程序，Apache 强烈推荐只使用一个 HttpClient 的实例。例如：

```
private static HttpClient httpClient = null;

private static synchronized HttpClient getHttpClient() {
    if(httpClient == null) {
        final HttpParams httpParams = new BasicHttpParams();
        httpClient = new DefaultHttpClient(httpParams);
    }

    return httpClient;
}
```

DefaultHttpClient 是常用的一个用于实现 HttpClient 接口的子类。HttpClietnt 定义的主要抽象方法就是 execute()，下面两种是常用的方法。

- HttpResponse execute (HttpUriRequest request)：通过 HttpUriRequest 对象的执行来返回一个 HttpResponse 对象。
- HttpResponse execute (HttpUriRequest request, HttpContext context)：通过 HttpUriRequest 对象和 HttpContext 对象的执行来返回一个 HttpResponse 对象。

Apache 对网络连接进行了管理，对于已经和服务端建立了连接的应用来说，再次调用 HttpClient 进行网络数据传输时，不必重新建立新连接，而可以重用已经建立的连接。客户端程序员不需要做任何配置，这样无疑可以减少开销，提升速度。

注意：Apache 的连接管理并不会主动释放建立的连接，需要程序员在不用的时候手动关闭连接。

5. 连接参数的设置

HttpParams 保存 Http 请求设定的参数对象。另外，还有 HttpConnectionParams 类，主要提供对 Http 连接参数(如连接超时时间等)进行设定的方法，例如：

```
private static synchronized HttpClient getHttpClient() {
    if(httpClient == null) {
        final HttpParams httpParams = new BasicHttpParams();

        //timeout: get connections from connection pool
        ConnManagerParams.setTimeout(httpParams, 1000);
        //timeout: connect to the server
        HttpConnectionParams.setConnectionTimeout(httpParams,
```

```
                DEFAULT_SOCKET_TIMEOUT);
        //timeout: transfer data from server
        HttpConnectionParams.setSoTimeout(httpParams,
                DEFAULT_SOCKET_TIMEOUT);

        //set max connections per host
        ConnManagerParams.setMaxConnectionsPerRoute(httpParams,
                new ConnPerRouteBean(DEFAULT_HOST_CONNECTIONS));
        //set max total connections
        ConnManagerParams.setMaxTotalConnections(httpParams,
                DEFAULT_MAX_CONNECTIONS);

        //use expect-continue handshake
        HttpProtocolParams.setUseExpectContinue(httpParams, true);
        //disable stale check
        HttpConnectionParams.setStaleCheckingEnabled(httpParams, false);

        HttpProtocolParams.setVersion(httpParams, HttpVersion.HTTP_1_1);
        HttpProtocolParams.setContentCharset(httpParams, HTTP.UTF_8);

        HttpClientParams.setRedirecting(httpParams, false);

        //set user agent
        String userAgent = "Mozilla/5.0 (Windows; U; Windows NT 5.1; zh-CN;
                rv:1.9.2) Gecko/20100115 Firefox/3.6";
        HttpProtocolParams.setUserAgent(httpParams, userAgent);

        //disable Nagle algorithm
        HttpConnectionParams.setTcpNoDelay(httpParams, true);

        HttpConnectionParams.setSocketBufferSize(httpParams,
                DEFAULT_SOCKET_BUFFER_SIZE);

        //scheme: http and https
        SchemeRegistry schemeRegistry = new SchemeRegistry();
        schemeRegistry.register(new Scheme("http",
                PlainSocketFactory.getSocketFactory(), 80));
        schemeRegistry.register(new Scheme("https",
                SSLSocketFactory.getSocketFactory(), 443));

        ClientConnectionManager manager = new ThreadSafeClientConnManager(
                httpParams, schemeRegistry);
        httpClient = new DefaultHttpClient(manager, httpParams);
    }

    return httpClient;
}
```

06～09 行通过 setTimeout()和 setConnectionTimeout()方法进行超时设置，让连接在超过时间后自动失效，释放占用资源。

注意：设置连接超时和请求超时，以下两个超时的意义不同。

- ConnManagerParams. setTimeout()：ConnectionManager 管理的连接池中取出连接的超时时间。
- HttpConnectionParams. setConnectionTimeout()：通过网络与服务器建立连接的超时时间。Httpclient 包中通过一个异步线程去创建与服务器的 Socket 连接，这就是该 Socket 连接的超时时间。

15～19 行的作用是连接数限制。配置每台主机最多连接数和连接池中的最多连接总数，对连接数量进行限制。其中，DEFAULT_HOST_CONNECTIONS 和 DEFAULT_MAX_CONNECTIONS 是由客户端程序员根据需要而设置的。

22 行的作用是持续握手验证。在认证系统或其他可能遭到服务器拒绝应答的情况下(如登录失败)，如果发送整个请求体，则会大大降低效率。此时，可以先发送部分请求(如只发送请求头)进行试探，如果服务器愿意接收，则继续发送请求体。

24 行的作用是关闭旧连接检查的配置。HttpClient 为了提升性能，默认采用了“重用连接”机制，即在有传输数据需求时，会首先检查连接池中是否有可供重用的连接，如果有，则会重用连接。同时，为了确保该“被重用”的连接确实有效，会在重用之前对其进行有效性检查。这个检查大概会花费 15～30 毫秒。关闭该检查举措，会稍微提升传输速度，但也可能出现“旧连接”过久而被服务器端关闭，从而出现 I/O 异常。

6. 多线程安全管理

如果应用程序采用了多线程进行网络访问，则应该使用 Apache 封装好的线程安全管理类 ThreadSafeClientConnManager 来进行管理，这样能够更有效且更安全地管理多线程和连接池中的连接。例如：

```
ClientConnectionManager manager = new ThreadSafeClientConnManager(
        httpParams, schemeRegistry);
httpClient = new DefaultHttpClient(manager, httpParams);
```

HTTP 连接是复杂的、有状态的、线程不安全的，对象需要正确的管理以便正确地执行功能。HTTP 连接在同一时间仅仅只能由一个执行线程来使用。HttpClient 采用一个特殊实体来管理访问 HTTP 的连接，这被称为 HTTP 连接管理器，代表了 ClientConnectionManager 接口。一个 HTTP 连接管理器的目的是作为工厂服务于新的 HTTP 连接，管理持久连接和同步访问持久连接来确保同一时间仅有一个线程可以访问一个连接。

下面介绍一个使用 Application 结合 HttpClient 单实例模式实现线程安全的 HttpClient 功能。

新建 HTTPClientApplication，代码如下：

```
public class HTTPClientApplication extends Application {
    private HttpClient mHttpClient = null;
```

```
    private static final String CHARSET = HTTP.UTF_8;

    @Override
    public void onCreate() {
        super.onCreate();
        mHttpClient = this.createHttpClient();
    }

    @Override
    public void onTerminate() {
        super.onTerminate();
        this.shutdownHttpClient();
    }

    @Override
    public void onLowMemory() {
        super.onLowMemory();
        this.shutdownHttpClient();
    }

    /**
     * 创建 HttpClient 实例
     * @return
     */
    private HttpClient createHttpClient(){
        HttpParams params = new BasicHttpParams();
        HttpProtocolParams.setVersion(params, HttpVersion.HTTP_1_1);
        HttpProtocolParams.setContentCharset(params, CHARSET);
        HttpProtocolParams.setUseExpectContinue(params, true);
        ConnManagerParams.setTimeout(params, 1000);
        HttpConnectionParams.setConnectionTimeout(params, 2000);
        HttpConnectionParams.setSoTimeout(params, 4000);
        SchemeRegistry schReg = new SchemeRegistry();
        schReg.register(new Scheme("http",
                PlainSocketFactory.getSocketFactory(), 80));
        schReg.register(new Scheme("https",
                SSLSocketFactory.getSocketFactory(), 443));

        //使用线程安全的连接管理来创建 HttpClient
        ClientConnectionManager conMgr = new
                ThreadSafeClientConnManager(params, schReg);
        HttpClient client = new DefaultHttpClient(conMgr, params);
        return client;
    }

    private void shutdownHttpClient(){
        if(mHttpClient != null && mHttpClient.getConnectionManager() != null){
            mHttpClient.getConnectionManager().shutdown();
        }
```

```
    }

    public HttpClient getHttpClient(){
        return mHttpClient;
    }

}
```

Application 和 Activity、Service 一样是 Android 框架的一个系统组件，当 Android 程序启动时，系统会创建一个 Application 对象，用来存储系统的一些信息。Android 系统会为每个程序运行时创建一个 Application 类的对象且仅创建一个，所以 Application 可以说是单实例模式的一个类，且 Application 对象的生命周期是整个程序中最长的，它的生命周期就等于这个程序的生命周期。因为它是全局的、单例的，所以在不同的 Activity、Service 中获得的对象都是同一个对象。因此，可以通过 Application 来进行一些数据传递、数据共享、数据缓存等操作。

继承自 Application 的类需要在 AndroidManifest. xml 的 application 标签中进行注册，例如：

```
<application android:name = "cn.njcit.project05.meta.HttpClientApplication"
    android:icon = "@drawable/icon"
    android:label = "@string/app_name">
```

使用这个 HTTPClientApplication 的方法是：在 Activity 的 onCreate()生命周期方法中，通过“mHTTPClientApplication = (HTTPClientApplication) getApplication();”和 mHTTPClientApplication. getHttpClient()就可以得到 HttpClient 的实例。可以看到，这个 HttpClient 实例在低内存和应用退出时关闭连接管理器，释放资源。

5.3.2 HttpGet

与 HttpURLConnection 相同，HttpClient 也存在 GET 和 POST 两种方式。

在 HttpClient 中，可以使用 HttpGet 对象来通过 GET 方式进行数据请求操作，当获得 HttpGet 对象后就可以使用 HttpClient 的 execute()方法来向服务器发送请求。在发送的 GET 请求被服务器响应后，会返回一个 HttpResponse 响应对象，利用这个响应的对象就能够获得响应回来的状态码，如 200、400、401 等。

使用 HttpGet 需要以下 6 个步骤：

① 创建 HttpClient 的实例。

② 创建连接方法的 HttpGet 实例，并在构造方法中传入待连接的地址。对于发送请求的参数，GET 和 POST 使用的方式不同，GET 方式可以使用拼接字符串的方式，把参数拼接在 URL 结尾；POST 方式需要使用 setEntity(HttpEntity entity)方法来设置请求参数。

③ 调用第一步中创建好的 HttpClient 实例的 execute()方法来执行第二步中创建好的 HttpGet 实例。

④ 读取返回的 HttpResponse 实例，通过 HttpResponse 接口的 getEntity()方法返回

响应信息,并进行处理。

⑤ 释放连接。无论执行方法是否成功,都必须释放连接。

⑥ 对得到后的内容进行处理。

下面的代码根据以上的步骤实现了使用 GET 方法从网络获取图像的功能:

```
01 public static Bitmap sendGetResquest(String path) {
02     Bitmap bitmap = null;
03     HttpGet httpGet = new HttpGet(path);
04     HttpClient httpClient = new DefaultHttpClient();
05
06     try {
07         HttpResponse httpResponse = httpClient.execute(httpGet);
08         int reponseCode = httpResponse.getStatusLine().getStatusCode();
09         if(reponseCode == HttpStatus.SC_OK) {
10             InputStream inputStream = httpResponse.getEntity().getContent();
11             bitmap = BitmapFactory.decodeStream(inputStream);
12             inputStream.close();
13         }
14     } catch (ClientProtocolException e) {
15         e.printStackTrace();
16     } catch (IOException e) {
17         e.printStackTrace();
18     }
19     return bitmap;
20 }
```

07 行的 HttpResponse 接口里定义了一系列的 set、get 方法。举例如下。

- HttpEntity getEntity():得到一个 HttpEntity 对象。
- StatusLine getStatusLine():得到一个 StatusLine(也就是 HTTP 协议中的状态行。HTPP 状态行由三部分组成,即 HTTP 协议版本、服务器发回的响应状态代码、状态码的文本描述)接口的实例对象。例如:

```
HttpResponse response = new BasicHttpResponse(HttpVersion.HTTP_1_1,
        HttpStatus.SC_OK, "OK");
response.getProtocolVersion();                          //HTTP/1.1
response.getStatusLine().getStatusCode();               //200
response.getStatusLine().getReasonPhrase();             //OK
response.getStatusLine().toString();                    //HTTP/1.1 200 OK
```

- Locale getLocale():得到 Locale 对象。

10 行通过 getEntity()得到一个 HttpEntity 对象,HttpEntity 是一个接口,通过 getContent()方法得到一个输入流对象 InputStream,可以用这个流来操作文件(例如保存文件到 SD 卡中)。

注意,当需要传输大量数据时,不应使用字符串或者字节数组,因为它们会将数据缓存至内存。当数据过多,尤其在多线程情况下,很容易造成内存溢出(Out Of Memory,

OOM)。而 HttpClient 能够有效处理实体流。这些"流"不会缓存至内存,而是直接进行数据传输。采用"请求流/响应流"的方式进行传输,可以减少内存占用,降低内存溢出的风险。例如:

```
//Get method: getResponseBodyAsStream()
//not use getResponseBody(), or getResponseBodyAsString()
GetMethod httpGet = new GetMethod(url);
InputStream inputStream = httpGet.getResponseBodyAsStream();
//Post method: getResponseBodyAsStream()
PostMethod httpPost = new PostMethod(url);
InputStream inputStream = httpPost.getResponseBodyAsStream();
```

5.3.3 HttpPost

POST 方法用来向目的服务器发出请求,要求它接收被附在请求后的实体,并把它当作请求队列中请求 URI 所指定资源的附加新子项。POST 被设计成用统一的方法实现下列功能:

- 对现有资源的注释。
- 向电子公告栏、新闻组、邮件列表或类似讨论组发送消息。
- 提交数据块,如将表单的结果提交给数据处理过程。
- 通过附加操作来扩展数据库。

POST 方法与 GET 方法的使用步骤大体相同。当使用 POST 方式时,可以使用 HttpPost 类来进行操作。当获取了 HttpPost 对象后,就需要向这个请求体传入键值对,这个键值对可以使用 NameValuePair 对象来进行构造,然后再使用 HttpRequest 对象最终构造的请求体,最后使用 HttpClient 的 execute()方法来发送请求,并在得到响应后返回一个 HttpResponse 对象。

下面是一个 HttpPost 请求示例:

```
public static String sendPostResquest(String path, Map<String,
        String> params, String encoding) {
    List<NameValuePair> list = new ArrayList<NameValuePair>();

    if((params != null) && !params.isEmpty()) {
        for(Map.Entry<String, String> param : params.entrySet()) {
            list.add(new BasicNameValuePair(param.getKey(),
                    param.getValue()));
        }
    }

    try {
        UrlEncodedFormEntity entity = new UrlEncodedFormEntity(list,
                encoding);
        HttpPost httpPost = new HttpPost(path);
        httpPost.setEntity(entity);
        HttpClient client = new DefaultHttpClient();
```

```
18          HttpResponse httpResponse = client.execute(httpPost);
19          int reponseCode = httpResponse.getStatusLine().getStatusCode();
20          if(reponseCode == HttpStatus.SC_OK) {
21              String resultData = EntityUtils.toString(
22                      httpResponse.getEntity());
23              return resultData;
24          }
25      } catch (IOException e) {
26          e.printStackTrace();
27      }
28      return "";
29  }
```

04 行的 NameValuePair 接口是一个简单封闭的键值对,只提供了一个 getName()和一个 getValue 方法。主要用到的实现类是 BasicNameVaulePair。

21 行的 EntityUtils 是一个 final 类,专门用于处理 HttpEntity,常用方法包括以下几种。

- String getContentCharSet (HttpEntity entity):设置 HttpEntity 对象的 ContentCharset。
- byte[] toByteArray (HttpEntity entity):将 HttpClient 转换成一个字节数组。
- String toString (HttpEntity entity, String defaultCharset):通过指定的编码方式取得 HttpEntity 里字符串的内容。
- String toString (HttpEntity entity):取得 HttpEntity 里字符串的内容。

5.3.4 AndroidHttpClient

android.net.http.AndroidHttpClient 类是 Google 对 Apache 中 HttpClient 的一个封装,属于 Android 原生的 Http 访问工具类。

AndroidHttpClient 对 DefaultHttpClient 做了一些改进,使其更易用于 Android 项目,包括以下内容:

- 关掉过期检查,自连接可以打破所有的时间限制。
- 可以设置 ConnectionTimeOut(连接超时)和 SoTimeout(读取数据超时)。
- 关掉重定向。
- 将一个 Session 缓冲用于 SSL Sockets。
- 如果服务器支持,使用 gzip 压缩方式用于在服务端和客户端之间传递数据。
- 默认情况下不保留 Cookie。

使用 AndroidHttpClient 的方式和 DefaultHttpClient 差不多。有一点需要注意的是,AndroidHttpClient 是一个 final 类,也没有公开的构造方法,所以无法使用 new 的形式对其进行实例化,必须使用 AndroidHttpClient.newInstance()方法获得 AndroidHttpClient 对象。

下面的代码是 Android SDK 自带的通过 AndroidHttpClient 下载图像的示例:

```
class BitmapDownloaderTask extends AsyncTask<String, Void, Bitmap> {
    private static final int IO_BUFFER_SIZE = 4 * 1024;
```

```
    private String url;
    private final WeakReference<ImageView> imageViewReference;

    public BitmapDownloaderTask(ImageView imageView) {
        imageViewReference = new WeakReference<ImageView>(imageView);
    }

    /**
     * Actual download method.
     */
    @Override
    protected Bitmap doInBackground(String... params) {
        final AndroidHttpClient client =
                AndroidHttpClient.newInstance("Android");
        url = params[0];
        final HttpGet getRequest = new HttpGet(url);
        String cookie = params[1];
        if (cookie != null) {
            getRequest.setHeader("cookie", cookie);
        }

        try {
            HttpResponse response = client.execute(getRequest);
            final int statusCode = response.getStatusLine().getStatusCode();
            if (statusCode != HttpStatus.SC_OK) {
                Log.w("ImageDownloader", "Error " + statusCode +
                        " while retrieving bitmap from " + url);
                return null;
            }

            final HttpEntity entity = response.getEntity();
            if (entity != null) {
                InputStream inputStream = null;
                OutputStream outputStream = null;
                try {
                    inputStream = entity.getContent();
                    final ByteArrayOutputStream dataStream =
                            new ByteArrayOutputStream();
                    outputStream = new BufferedOutputStream(dataStream,
                            IO_BUFFER_SIZE);
                    copy(inputStream, outputStream);
                    outputStream.flush();

                    final byte[] data = dataStream.toByteArray();
                    final Bitmap bitmap = BitmapFactory.decodeByteArray(
                            data, 0, data.length);

                    //FIXME : Should use BitmapFactory
                    //.decodeStream(inputStream) instead.
```

```
                    //final Bitmap bitmap = BitmapFactory
                    //.decodeStream(inputStream);

                    return bitmap;

                } finally {
                    if (inputStream != null) {
                        inputStream.close();
                    }
                    if (outputStream != null) {
                        outputStream.close();
                    }
                    entity.consumeContent();
                }
            }
        } catch (IOException e) {
            getRequest.abort();
            Log.w(LOG_TAG, "I/O error while retrieving bitmap from " + url, e);
        } catch (IllegalStateException e) {
            getRequest.abort();
            Log.w(LOG_TAG, "Incorrect URL: " + url);
        } catch (Exception e) {
            getRequest.abort();
            Log.w(LOG_TAG, "Error while retrieving bitmap from " + url, e);
        } finally {
            if (client != null) {
                client.close();
            }
        }
        return null;
    }

    /**
     * Once the image is downloaded, associates it to the imageView
     */
    @Override
    protected void onPostExecute(Bitmap bitmap) {
        if (isCancelled()) {
            bitmap = null;
        }

        //Add bitmap to cache
        if (bitmap != null) {
            synchronized (sHardBitmapCache) {
                sHardBitmapCache.put(url, bitmap);
            }
        }

        if (imageViewReference != null) {
```

```
            ImageView imageView = imageViewReference.get();
            BitmapDownloaderTask bitmapDownloaderTask =
                    getBitmapDownloaderTask(imageView);
            //Change bitmap only if this process is still associated with it
            if (this == bitmapDownloaderTask) {
                imageView.setImageBitmap(bitmap);
            }
        }
    }

    public void copy(InputStream in, OutputStream out) throws IOException {
        byte[] b = new byte[IO_BUFFER_SIZE];
        int read;
        while ((read = in.read(b)) != -1) {
            out.write(b, 0, read);
        }
    }
}
```

5.4 Http 连接框架

目前,Android 平台上基于 Http 的连接框架有很多种,本节主要介绍应用比较广泛的 android-async-http 框架和 Volley 框架。

5.4.1 android-async-http 框架

android-async-http① 是 Android 上的一个异步 HTTP 客户端开发包,它构建在 Apache HttpClient 库之上,所有的请求都在应用程序的主线程(UI 线程)之外执行,但任何回调的逻辑都使用 Android 的 Handler 消息传递机制在同一线程中被执行。

1. 功能特性

android-async-http 提供的主要功能包括:

- 进行异步 HTTP 请求,并通过匿名内部类处理回调结果。
- HTTP 请求发生在 UI 线程之外,不会阻塞 UI 操作。
- 通过线程池处理并发请求。
- GET/POST 参数构造通过 RequestParams 类完成。
- 处理文件的上传、下载。
- 响应结果自动打包成 JSON 格式。
- 应用程序开销少,仅仅 25KB 就可以满足所有要求。
- 能处理环行和相对重定向。
- 自动智能请求重试,优化了质量不一的移动连接。

① 项目主页:http://loopj.com/android-async-http/。

- 自动的 GZIP 响应解码支持超快速的请求。
- 二进制文件通过 BinaryHttpResponseHandler 下载。
- 内置响应通过 JsonHttpResponseHandler 解析成 JSON。
- 支持 SAX 解析器。
- 持久化 Cookie 存储、保存 Cookie 到应用程序的 SharedPreferences。
- 支持各种语言和 content 编码(不仅仅是 UTF-8)。

2. 核心类

(1) AsyncHttpRequest

AsyncHttpRequest 继承自 Runnable,基于线程的子类,用于异步请求,通过 AsyncHttpResponseHandler 回调。

(2) AsyncHttpResponseHandler

AsyncHttpResponseHandler 接收请求结果,一般重写 onSuccess()及 onFailure()方法来接收请求成功或失败的消息,还有 onStart()、onFinish()等消息。

(3) TextHttpResponseHandler

TextHttpResponseHandler 继承自 AsyncHttpResponseHandler,只是重写了 AsyncHttpResponseHandler 的 onSuccess()及 onFailure()方法,将请求结果由 byte 数组转换为 String。

(4) JsonHttpResponseHandler

JsonHttpResponseHandler 继承自 TextHttpResponseHandler,同样是重写 onSuccess()及 onFailure()方法,将请求结果由 String 转换为 JSONObject 或 JSONArray。

(5) BaseJsonHttpResponseHandler

BaseJsonHttpResponseHandler 继承自 TextHttpResponseHandler,是一个泛型类,提供了 parseResponse()方法,子类需要提供实现,将请求结果解析成需要的类型,子类可以灵活地使用解析方法,可以直接原始解析,或使用 gson 等。

(6) RequestParams

RequestParams 为请求参数,可以添加普通的字符串参数,也可添加 File,使用 InputStream 上传文件。

(7) AsyncHttpClient

AsyncHttpClient 为核心类,使用 HttpClient 执行网络请求,提供了 GET、PUT、POST、DELETE、HEAD 等请求方法,使用起来很简单,只需以 URL 及 RequestParams 调用相应的方法即可,还可以选择性地传入 Context,用于取消 Content 相关的请求,同时必须提供 ResponseHandlerInterface(AsyncHttpResponseHandler 继承自 ResponseHandler Interface)的实现类,一般为 AsyncHttpResponseHandler 的子类,AsyncHttpClient 内部有一个线程池,当使用 AsyncHttpClient 执行网络请求时,最终都会调用 sendRequest()方法,在这个方法内部将请求参数封装成 AsyncHttpRequest 并交由内部的线程池执行。

(8) SyncHttpClient

SyncHttpClient 继承自 AsyncHttpClient,同步执行网络请求,AsyncHttpClient 把请求封装成 AsyncHttpRequest 后提交至线程池,SyncHttpClient 把请求封装成 AsyncHttpRequest 后直接调用它的 run()方法。

3. 请求步骤

使用 android-async-http 的一般步骤如下。

① 创建一个 AsyncHttpClient。下面的代码是一个使用 AsyncHttpClient 的典型框架。

```
AsyncHttpClient client = new AsyncHttpClient();
client.get(httpUrl, new AsyncHttpResponseHandler() {

    @Override
    public void onStart() {
        //called before request is started
    }

    @Override
    public void onSuccess(int statusCode, Header[] headers,
            byte[] response) {
        //called when response HTTP status is "200 OK"
    }

    @Override
    public void onFailure(int statusCode, Header[] headers,
            byte[] errorResponse, Throwable e) {
        //called when response HTTP status is "4XX" (eg. 401, 403, 404)
    }

    @Override
    public void onRetry(int retryNo) {
        //called when request is retried
    }
});
```

② (可选的)通过 RequestParams 对象设置请求参数。例如：

```
RequestParams params = new RequestParams();
params.put("username", userName);
params.put("userpass", userPass);
```

③ 调用 AsyncHttpClient 的 get 或 post 等方法发起网络请求。例如：

```
mAsyncHttpClient.post(url, params, new AsyncHttpResponseHandler() {});
```

④ 所有的请求都在 sendRequest 中被封装为 AsyncHttpRequest，并添加到线程池执行。

⑤ 当请求被执行时(即 AsyncHttpRequest 的 run()方法)，执行 AsyncHttpRequest 的 makeRequestWithRetries()方法执行实际的请求，当请求失败时可以重试。并在请求开始、结束、成功或失败时向请求时传入的 ResponseHandlerInterface 实例发送消息。

⑥ 使用 AsyncHttpResponseHandler 的子类并调用其 onStart()、onSuccess()等方法返回请求结果。

下面的 downloadImage()方法演示了使用 android-async-http 框架的典型用法。

```
private void downloadImage(String url) {
    AsyncHttpClient client = new AsyncHttpClient();
    client.get(url, new AsyncHttpResponseHandler() {
        @Override
        public void onSuccess(int statusCode, Header[] headers,
                byte[] responseBody) {
            if (statusCode == 200) {
                BitmapFactory factory = new BitmapFactory();
                Bitmap bitmap = factory.decodeByteArray(responseBody, 0,
                        responseBody.length);
                imageView.setImageBitmap(bitmap);
            }
        }

        @Override
        public void onFailure(int statusCode, Header[] headers,
                byte[] responseBody, Throwable error) {
            error.printStackTrace();
        }
    });
}
```

5.4.2 Volley 框架

Volley[1] 是 Google 工程师 Ficus Kirpatrick 在 Gooogle I/O 2013 发布的一个处理和缓存网络请求的库，内部使用了 HttpURLConnection 和 HttpClient[2] 网络连接，能使网络通信更快、更简单、更健壮。

1. 功能特性

Volley 适合传输数据量不大，但通信频繁的网络操作，特别是针对 JSON 对象和图像加载的应用。

Volley 提供的主要特性包括：

- 封装了异步的 RESTful[3] 请求 API；
- 请求队列的优先级排序；
- 认证头部管理；
- 一个可扩展的架构，它使开发人员能够实现自定义的请求和响应处理机制；

① 源码下载地址：git clone https://android.googlesource.com/platform/frameworks/volley。

② Android 2.3 之前的版本，建议使用 Apache 的 HttpClient 来处理网络请求，2.3 之后的版本建议使用 HttpUrlConnection。博文参见：http://android-developers.blogspot.com/2011/09/androids-http-clients.html。

③ RESTful 是一种软件架构风格，提供了一组设计原则和约束条件，它主要用于客户端和服务器交互类的软件，基于这个风格设计的软件可以更简洁、更有层次、更易于实现缓存等机制。

- 能够使用外部 HTTP Client 库；
- 透明的缓存策略；
- UI 线程可更新；
- 提供了调试和跟踪工具，并可以在 UI 线程提示请求过程中出现的错误；
- 自定义的网络图像加载视图（如 NetworkImageView、ImageLoader 等）；
- Volley 可以绑定 Content 的生命周期，例如当 Activity 销毁时，Volley 可以取消响应的网络请求。

2. Volley 架构

Volley 使用线程池来作为基础结构，主要分为主线程、Cache 线程和 Network 线程。主线程和 Cache 线程都只有一个，而 NetworkDispatcher 线程可以有多个，这样能解决并行问题。图 5-1 的 Volley 体系框架节选自 Ficus Kirpatrick 在 Google 会议上的演讲。

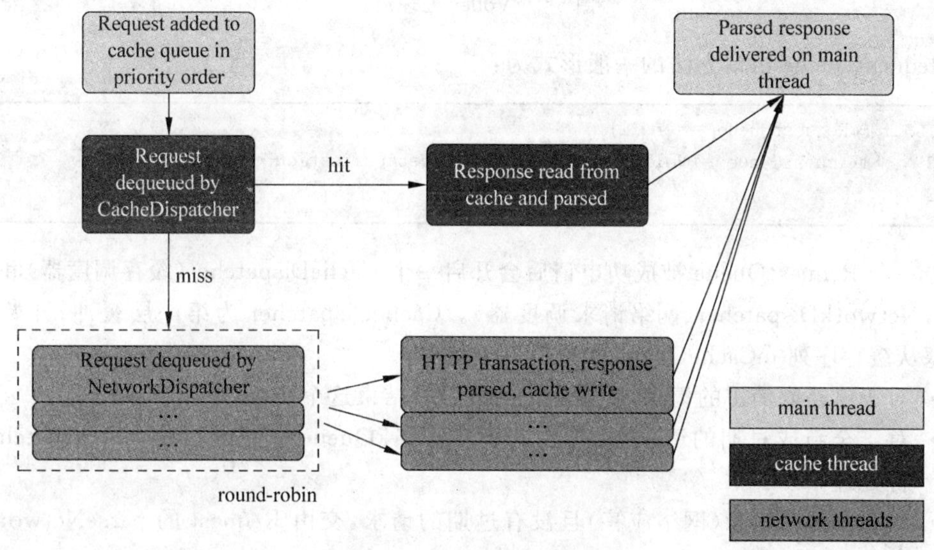

图 5-1 Volley 体系架构

3. Volley 处理流程

Volley 一个很大的特色，就是所有的网络请求无须开发者自己执行，而是在请求构造完成后传递到 Volley 的请求队列中，队列依次进行请求，开发者不用担心网络请求是否会冲突，是否会在主线程，从而提高了开发效率。

Volley 的基本处理流程如图 5-2 所示。

(1) 应用初始化 Volley。

Volley 是一个 Helper 类，主要负责创建请求队列，并且启动队列消息的循环。

(2) Volley 创建一个 RequestQueue、NetworkDispatcher 组及 Network。

RequestQueue 即一个 Request 队列，RequestQueue 会创建一个 ExecutorDelivery。

RequestQueue 是一个请求队列对象，它可以缓存所有的 HTTP 请求，然后按照一定的算法并发地发出这些请求。RequestQueue 内部的设计非常适合高并发应用，程序不必为每一次 HTTP 请求都创建一个 RequestQueue 对象（基本上在每一个需要和网络交互的 Activity 中创建一个 RequestQueue 对象就足够了）。

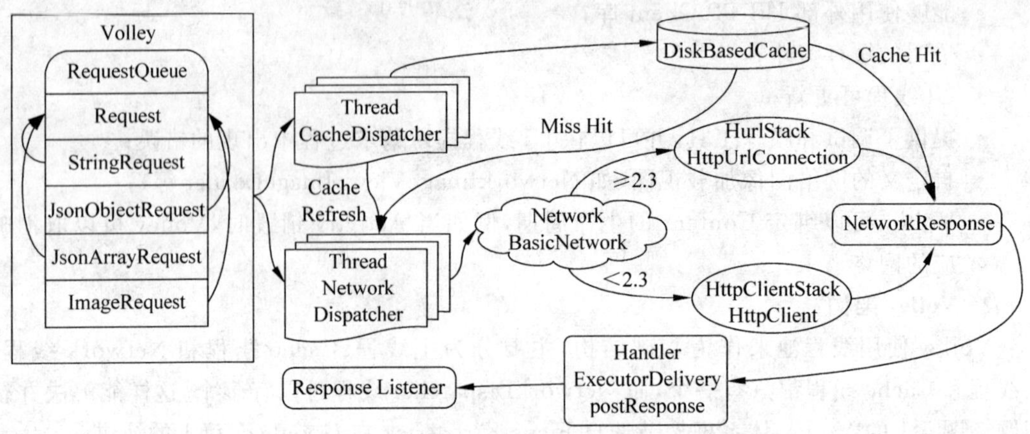

图 5-2　Volley 处理流程

RequestQueue 申请方法的一般形式为：

```
if (mRequestQueue == null) {
    mRequestQueue = Volley.newRequestQueue(getApplicationContext());
}
```

当一个 RequestQueue 被成功申请后会开启一个 CacheDispatcher(缓存调度器)和 4 个(默认)NetworkDispatcher(网络请求调度器)。CacheDispatcher 为第一层缓冲，开始工作后阻塞从缓存序列 mCacheQueue 中取得以下请求：

- 对于已经取消了的请求，直接标记为跳过并结束这个请求。
- 对于全新或过期的请求，直接丢入 mNetworkQueue 中交由 NetworkDispatcher 进行处理。
- 已获得缓存信息(网络应答)且没有过期的请求，交由 Request 的 parseNetworkResponse 进行解析，从而确定此应答是否成功。然后将请求和应答交由 Delivery 分发者进行处理，如果需要更新缓存，那么该请求还会被放入 mNetworkQueue 中。

(3) NetworkDispatcher 负责网络调度(实质是 Thread 线程池)，从 RequestQueue 中取 Request，通过 Network 加以执行。

Volley 提供了 JsonObjectRequest、JsonArrayRequest、StringRequest 等 Request 形式。

- JsonObjectRequest：返回 JSON 对象。
- JsonArrayRequest：返回 JsonArray。
- StringRequest：返回 String。

另外，可以通过继承 Request<T>来自定义 Request。

用户将 Request 添加到 RequestQueue 之后，还需做以下事情。

- 对于不需要缓存的请求(需要额外设置，默认是需要缓存)直接丢入 mNetworkQueue 交由 NetworkDispatcher 处理。
- 对于需要缓存的全新的请求，则加入 mCacheQueue 中给 CacheDispatcher 处理。
- 需要缓存但是缓存列表中已经存在了相同 URL 的请求，则放在 mWaitingQueue 中做暂时保管，待之前的请求结束后，再重新添加到 mCacheQueue 中。

下面是一个 StringRequest 的示例：

```
StringRequest req = new StringRequest(URL,
    new Response.Listener<String>() {
        @Override
        public void onResponse(String response) {
            VolleyLog.v("Response: %n %s", response);
        }
    },
    new Response.ErrorListener() {
        @Override
        public void onErrorResponse(VolleyError error) {
        VolleyLog.e("Error: ", error.getMessage());
    }
});
```

(4) Network 负责网络请求的处理，具体过程交给 HttpStack 处理。Network 封装了 HttpStack 的网络请求，PerformRequest 方法直接返回一个 NetworkResponse，处理请求异常并抛出 VolleyError。

HttpStack 分 HttpURLConnection 与 HttpClient 两种方式。HttpStack 负责解析网络请求，PerformRequest 返回一个 Apache 的 HttpResponse，确保返回的 Response 有 StatusLine。

ExecutorDelivery 负责处理请求结果，并与主线程进行交互。Delivery 实际上已经是对网络请求处理的最后一层了，在 Delivery 对请求处理之前，Request 已经对网络应答进行过解析，此时应答成功与否已经设定。而后 Delivery 会根据请求所获得的应答情况做以下的不同处理：

- 若应答成功，则触发 deliverResponse 方法，最终会触发开发者为 Request 设定的 Listener。
- 若应答失败，则触发 deliverError 方法，最终会触发开发者为 Request 设定的 ErrorListener。

(5) 取消 Request。

如果在结束 Activity 时，其中启动的网络请求还没有返回结果，需要手动取消所有或部分未完成的网络请求。Volley 里所有的请求结果会返回给主进程，如果在主进程里取消了某些请求，则这些请求将不会被返回给主线程。Volley 支持多种 Request 取消方式。

- 针对某些 Request 做取消操作，例如：

```
@Override
public void onStop() {
    for (Request<?> req : mRequestQueue) {
        req.cancel();
    }
}
```

• 取消这个队列里的所有请求,例如:

```
@Override
protected void onStop() {
    super.onStop();
    mRequestQueue.cancelAll(this);
}
```

• 根据 RequestFilter 或者 Tag 来终止某些请求,例如:

```
@Override
protected void onStop() {
    super.onStop();
    mRequestQueue.cancelAll( new RequestFilter() {});
}
```

4. 使用 Volley 异步加载图像

(1) ImageRequest

ImageRequest 能够处理单张图像,返回 Bitmap 对象。ImageRequest 的构造方法如下:

```
ImageRequest(String url, Response.Listener<Bitmap> listener, int maxWidth,
        int maxHeight, Config decodeConfig,
        Response.ErrorListener errorListener)
```

第一个参数就是图像的 URL 地址。第二个参数是图像请求成功的回调。第三、第四个参数分别用于指定允许图像最大的宽度和高度,如果指定的网络图像的宽度或高度大于这里的最大值,则会对图像进行压缩,指定成 0 就表示不管图像有多大,都不会进行压缩。第五个参数用于指定图像的颜色属性,Bitmap.Config 下的几个常量都可以在这里使用,其中 ARGB_8888 可以展示最好的颜色属性,每个图像像素占据 4 个字节大小,而 RGB_565 则表示每个图像像素占据 2 个字节大小。第六个参数是图像请求失败的回调。

下面是 ImageRequest 的使用示例:

```
singleImg = (ImageView)findViewById(R.id.volley_img_single_imgeview);
ImageRequest imgRequest = new ImageRequest(url,
        new Response.Listener<Bitmap>() {
            @Override
            public void onResponse(Bitmap arg0) {
                singleImg.setImageBitmap(arg0);
            }
        },
        maxWidth, maxHeight, Config.ARGB_8888,
        new ErrorListener(){
            @Override
            public void onErrorResponse(VolleyError arg0) {
        }
});
mRequestQueue.add(imgRequest);
```

(2) ImageLoader

ImageLoader 类(内部也是使用 ImageRequest 来实现)需要一个 Request 的实例以及一个 ImageCache 的实例。图像通过一个 URL 和一个 ImageListener 实例的 get()方法来加载。ImageLoader 会检查 ImageCache,如果缓存里没有图像,就会从网络上获取。因此,ImageLoader 明显要比 ImageRequest 更加高效,因为它不仅可以对图像进行缓存,还可以过滤掉重复的链接,避免重复发送请求。

Volley 的 ImageCache 接口允许使用 L1 缓存实现,用户可以根据需要自行定义。ImageCache 接口有两个方法,即 getBitmap(String url)和 putBitmap(String url, Bitmap bitmap)来实现缓存机制。例如:

```
RequestQueue mRequestQueue = Volley.newRequestQueue(this);
LruCache<String, Bitmap> mImageCache = new LruCache<String, Bitmap>(20);

ImageCache imageCache = new ImageCache() {
    @Override
    public void putBitmap(String key, Bitmap value) {
        mImageCache.put(key, value);
    }

    @Override
    public Bitmap getBitmap(String key) {
        return mImageCache.get(key);
    }
};

ImageLoader mImageLoader = new ImageLoader(mRequestQueue, imageCache);
ImageListener listener = ImageLoader.getImageListener(
        imageView, android.R.drawable.ic_menu_rotate,
        android.R.drawable.ic_delete);
mImageLoader.get(url, listener);
```

(3) NetworkImageView

NetworkImageView 继承自 ImageView,是 Volley 提供的一个全新的、简单加载图像的控件。下面是一个在布局中使用的示例:

```
<com.android.volley.toolbox.NetworkImageView
        android:id="@+id/network_image_view"
        android:layout_width="200dp"
        android:layout_height="200dp"
        android:layout_gravity="center_horizontal"
/>
```

在 Activity 中绑定了 NetworkImageView 对象后,可以设置一些相关的属性,例如:

```
networkImageView.setDefaultImageResId(R.drawable.default_image);
networkImageView.setErrorImageResId(R.drawable.failed_image);
networkImageView.setImageUrl(url, imageLoader);
```

setImageUrl()方法的第一个参数用于指定图像的 URL 地址,第二个参数则是前面创建好的 ImageLoader 对象(其实 NetworkImageView 控件和用 ImageLoader 加载图像的性质是一样的)。

5.5 习　题

1. 编程实现从网络下载一幅图像并保存到 SD 卡中。
2. 编程实现使用 Volley 从网络异步下载多幅图像并以列表样式显示。

第6章　Web应用编程

Web App 无须安装，对设备碎片化的适应能力优于 Native App，它只需要通过 HTML、CSS 和 JavaScript 就可以在任意移动浏览器中执行。随着 WebKit 浏览体验的升级，使得支持 WebKit 浏览内核的移动设备开发的 Web App 也有了类似 Native App 一样流畅的用户体验。

6.1　访问 Web 页面

在 Android 中，有两种浏览 Web 网页的方法。

6.1.1　通过 Intent 浏览 Web 页面

Android 中提供了 Intent 机制来协助应用之间的交互与通信，Intent 负责对应用中一次操作的动作、动作涉及的数据以及附加数据等信息进行描述。Android 则根据此 Intent 的描述，负责找到对应的组件，将 Intent 传递给调用的组件，并完成组件的调用。因此，Intent 相当于应用程序之间的通信网络，是对执行某个操作的一种抽象描述。

下面的代码演示了通过隐式 Intent 在浏览器中访问 Google 主页的方法：

```
01  Uri uri = Uri.parse("http://www.google.com");
02  Intent intent = new Intent(Intent.ACTION_VIEW, uri);
03  startActivity(intent);
```

01 行通过 Uri.parse()方法将表示网址的字符串格式化，02 行构建了一个隐式的 Intent，其中的 ACTION_VIEW 是设置 Intent 的动作，多用于处理未知文件类型的打开。03 行通过调用 startActivity()方法将 Intent 发布成一个任务，可以不用确切知道哪个应用程序组件将会去执行该任务，只要关心该任务是否被执行、是否按照要求完成就够了，剩下的细节由系统自动去完成。

当隐式的 Intent 被抛出后，系统在众多组件中根据 Intent 过滤器中的 action、datatype、uri 来寻找与其匹配的处理方法。如果存在多个结果，用户可以根据需要选择合适的处理方法。

最后，需要在 AndroidManifest.xml 中添加如下访问网络的权限：

```
<uses-permission android:name = "android.permission.INTERNET"/>
```

图 6-1 显示了通过隐式的 Intent 访问 Web 页面的效果。

图 6-1　AVD 中的浏览器

6.1.2　通过 WebView 浏览 Web 页面

WebView 是 Android 的 View 类的子类,可以在 Activity 布局中嵌入一个 WebView 组件来显示一个网页。下面介绍 WebView 的基本使用。

在项目的 Res/layout 中创建一个包含 WebView 控件的布局文件,代码如下:

```
<?xml version = "1.0" encoding = "utf - 8"?>
< WebView xmlns:android = "http://schemas.android.com/apk/res/android"
    android:id = "@ + id/webview"
    android:layout_width = "fill_parent"
    android:layout_height = "fill_parent" />
```

在 Activity 的 onCreate()生命周期方法中通过 setContentView()方法载入上面包含 WebView 的布局,并绑定其中的 WebView 控件,代码如下:

```
01  private WebView webview;
02  ...
03  webview = (WebView) findViewById(R.id.webview);
04  webview.loadUrl("http://www.google.com");
```

04 行使用 loadUrl()方法载入网页。

最后,也需要在 AndroidManifest.xml 中添加如下访问网络的权限:

```
< uses - permission android:name = "android.permission.INTERNET"/>
```

程序的运行结果类似图 6-1。

6.2　WebKit 与 WebView

本节介绍 WebKit 和 WebView 的基本知识。

6.2.1　WebKit 浏览器引擎

WebKit[①] 是一个开源的浏览器引擎，包含一个网页引擎 WebCore 和一个脚本引擎 JavaScriptCore。Android 平台的 Web 引擎框架采用了 WebKit 项目中的 WebCore 和 JavaScriptCore 部分，上层由 Java 语言封装，并且作为 API 提供给 Android 应用开发者，而底层使用 WebKit 核心库（WebCore 和 JSCore）进行网页排版。Android 平台架构的 WebKit 如图 6-2 所示。

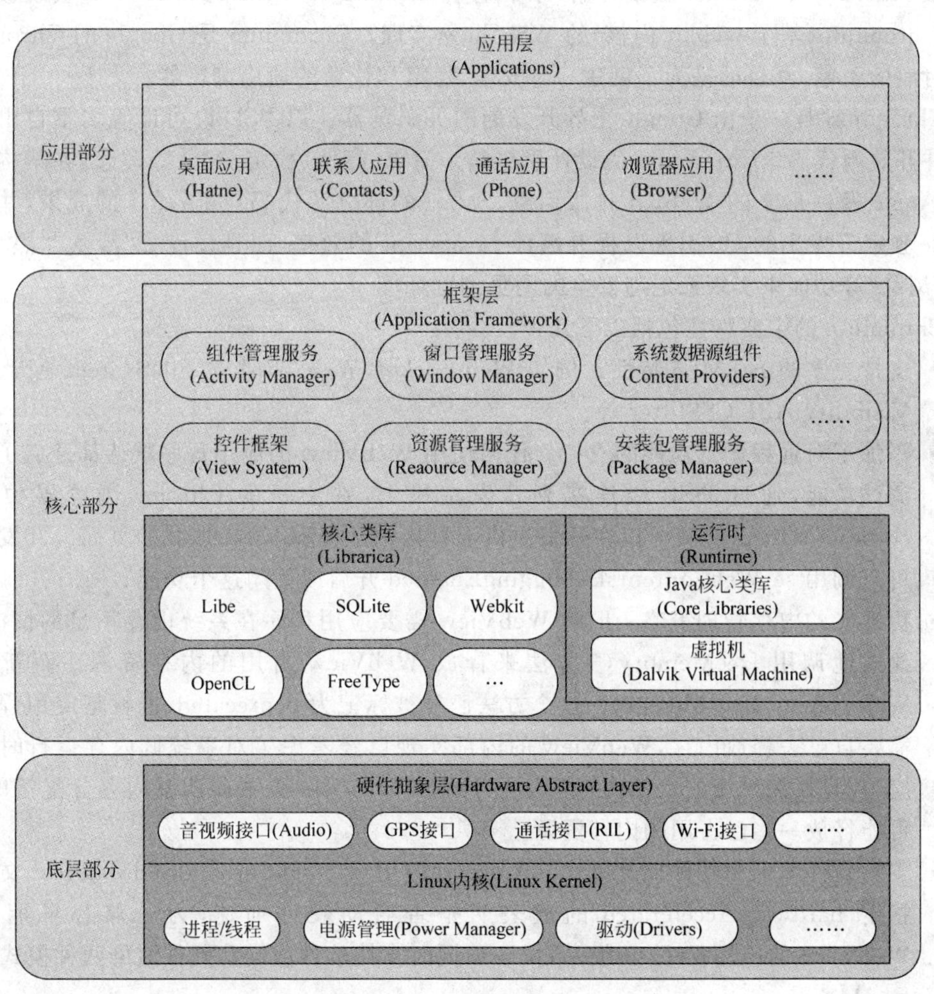

图 6-2　Android 平台架构

① Webkit 官网：http://www.webkit.org/。

Android 平台的 WebKit 由 Java 层和 WebKit 库两部分组成,Java 层负责与 Android 应用层进行通信,而 WebKit 类库负责实际的网页排版处理。Android SDK 中提供了 android. webkit. WebView 类,它继承自 View 类,为客户提供客户化浏览显示的功能,如果客户需要加入浏览器支持,可将该类的实例或者派生类的实例作为视图,调用 Activity 类的 setContentView()方法显示给用户。当客户代码中第一次生成 WebView 对象时,会初始化 WebKit 库(包括 Java 层和 C 层两个部分),之后用户可以操作 WebView 对象完成网络或者本地资源的访问。

WebKit 工作在两个线程中,一个是 UI 线程,也就是应用程序使用 WebView 所在的主线程;另一个是 WebCore 线程。WebView 的消息处理,主要是 UI 线程和 WebCore 线程的交互。一部分 UI 线程向 WebCore 发送命令操作的消息,例如 LOAD_URL;另一部分是来自 UI 的触摸消息。

从 Android 4.4 Kitkat 版本开始,原本基于 Android WebKit 的 WebView 实现被换成基于 Chromium(基于 webkit 内核)的 WebView 实现。Chromium 是 Google 的 Chrome 浏览器背后的引擎,其目的是为了创建一个安全、稳定和快速的通用浏览器。

Chromium 是一个由 Google 主导开发的网页浏览器。以 BSD 许可证等多重自由版权发行并开放源代码。Chromium 的设计思想基于简单、高速、稳定、安全等理念,在架构上使用了 Apple 发展出来的 WebKit 排版引擎、Safari 的部分源代码与 Firefox 的成果,并采用 Google 独家开发出的 V8 引擎以提升编译 JavaScript 的效率,而且设计了"沙盒""黑名单""无痕浏览"等功能来实现稳定与安全的网页浏览环境。

Chromium 的主要特性包括以下几种。

- 支持更多的 HTML5 特性。例如 Web Socket、Web Worker、FileSystem API、Page Visibility API、CSSfilter 等。
- 添加了对远程调试功能的支持。任何使用 WebView 的应用程序默认都开启了远程调试功能,通过 USB 线连接到开发主机上,在主机的 Chrome 浏览器中输入 chrome://inspect 就可以在开发主机上直接调试 WebView 加载的页面。开发者也可以调用 setWebContentsDebuggingEnabled 开启或关闭这个功能。
- 更智能的内存管理策略。旧版 WebView 需要应用程序在系统内存不足的情况下,显式地调用 freeMemory()方法来释放 WebView 占用的内存资源。而新版的 WebView 中,freeMemory()这个方法已经被标记为 deprecated,也就是应用程序不需要自己去释放内存,WebView 的内部实现已经考虑了对系统低内存运行时的响应,一旦发现系统处于低内存的运行状态,WebView 会主动调整内存分配策略,尽可能释放一些已占用的内存资源。
- 支持软件渲染和硬件加速模式。Android 应用程序可以在 AndroidManifest 文件中指定 hardwareAccelerated 的值表明是否启动硬件加速,为了兼容早期使用 WebView 的应用程序,新版 WebView 同时支持软件渲染和硬件渲染两种模式。

6.2.2 WebView 核心方法

WebView 是一种嵌入式的编程接口,能够提供 Java 接口给开发者来使用该模块渲染网页。现在的 WebView 只是一个接口类,通过一些内部设计的改变,其具体的实现可以在之前的 Android WebKit 和 Chromium 之间进行切换。新的 Chromium 实现专注于提供一致性的接口(为了兼容以前的应用),而内部的渲染引擎改为使用基于 Blink/Content 内核的引擎,这种实现不管是从功能上还是性能来讲,都带来巨大的提升。

WebView 的核心方法包括以下几种。

1. 页面加载方法

(1) public void loadData(String data, String mimeType, String encoding)

作用:加载指定的 data 数据。

参数说明如下。

- data:String 类型的数据,可以通过 base64 编码实现。
- mineType:data 数据的 MIME 类型,如"text/html"。
- encoding:data 数据的编码格式。

说明:

① Javascript 有同源限制,同源策略限制了一个源中加载文本或者脚本与来自其他源中的数据交互方式。要避免这种限制可以使用 loadDataWithBaseURL()方法。

② encoding 参数指定 data 参数是否为 base64 或者 URL 编码,如果 data 是 base64 编码,那么 encoding 必须填写为 base64。

(2) public void loadUrl(String url)

作用:加载指定 url 的网页内容。

(3) public void loadUrl(String url, Map<String, String> additionalHttpHeaders)

作用:加载指定 url 并携带 http header 数据。

WebView 的 loadUrl()方法除了可以加载网页,还可以实现如下功能。

① 打开本包内 asset 目录下的 index.html 文件。

```
webView.loadUrl(" file:///android_asset/index.html ");
```

② 打开本地 SD 卡内的 index.html 文件。

```
webView.loadUrl("content://com.android.htmlfileprovider/sdcard/" +
        "index.html");
```

③ 直接载入 HTML 字符串。

```
String htmlString = "<h1>Title</h1><p>This is HTML text<br />" +
        "<i>Formatted in italics</i><br />Anothor Line</p>";
webView.loadData(htmlString, "text/html", "utf-8");
```

(4) public void reload()

作用:页面所有资源都会重新加载。

(5) public void stopLoading()

作用:停止加载页面。

2. 页面导航方法

(1) public void goBack()、public void goForward()、public void goBackOrForward(int steps)

作用：3 个方法以当前的 index 为起始点前进或者后退到历史记录中指定的 steps，如果 steps 为负数，则后退；正数则前进。

(2) public boolean canGoForward()、public boolean canGoBack()

作用：两个方法用于设置是否允许页面前进或后退。

3. JavaScript 操作方法

(1) public void addJavascriptInterface(Object object, String name)

作用：当网页需要和 App 进行交互时，可以注入 Java 对象提供给 JavaScritp 调用。Java 对象提供相应的方法供 JavaScript 使用。

说明：Android 4.2 以下版本使用这个 API 会涉及 JavaScript 的安全问题，Javascript 可以通过反射这个 Java 对象的相关类进行攻击。

Android 4.2 及其以上系统需要给提供 JavaScript 调用的方法前加入一个注解为 @JavaScriptInterface。在虚拟机当中 Javascript 调用 Java 方法时会检测这个注解，如果方法被标识，则 Javascript 可以成功调用这个 Java 方法，否则调用不成功。

例如：

```
class JsObject {
    @JavascriptInterface
    public String toString() { return "injectedObject"; }
}
...
webView.addJavascriptInterface(new JsObject(), "injectedObject");
```

(2) public void evaluateJavascript(String script, ValueCallback<String> resultCallback)

作用：这个方法在 Android 4.4 系统中引入，因此只能在 Android 4.4 系统中才能使用，其用于在当前页面显示上下文中异步执行的 Javascript 代码。

说明：这个方法必须在 UI 线程调用，这个方法的回调也会在 UI 线程执行。Android 4.4 以下通过 WebView 提供的 loadUrl 方法调用，如：

```
webView.loadUrl("javascript:alert(injectedObject.toString())");
```

其中 javascript 是执行 Javascript 代码的标识，后面是 Javascript 语句。

(3) public void removeJavascriptInterface(String name)

作用：删除 addJavascripInterface()时对 WebView 注入 Java 对象。此方法在不同 Android 系统的 WebView 中会有问题，会存在失效情况。

4. 网页查找功能

(1) public void findAllAsync(String find)

作用：异步执行查找网页内包含的字符并设置为高亮显示，查找结果会回调。

(2) public int findAll(String find)

说明：这个 API 在 Android 4.1 中已经被删除。Android 4.1 以上系统使用 findAllAsync()

方法。

(3) public void findNext(boolean forward)

作用：查找下一个匹配的字符。

例如：

```
public class TestFindListener implements
        android.webkit.WebView.FindListener {
    private FindListener mFindListener;

    public TestFindListener(FindListener findListener) {
        mFindListener = findListener;
    }

    @Override
    public void onFindResultReceived(int activeMatchOrdinal,
            int numberOfMatches, boolean isDoneCounting) {
        mFindListener.onFindResultReceived(activeMatchOrdinal,
                numberOfMatches, isDoneCounting);
    }
}

public void findAllAsync(String searchString) {
    if (android.os.Build.VERSION_CODES.JELLY_BEAN <= Build.VERSION.SDK_INT)
        mWebView.findAllAsync(searchString);
    else {
        int number = mWebView.findAll(searchString);
        if (mIKFindListener != null)
            mIKFindListener.onFindResultReceived(number);
            fixedFindAllHighLight();
        }
    }

mWebView.findNext(forward);
```

5. 数据清除部分

(1) public void clearCache(boolean includeDiskFiles)

作用：清除网页访问留下的缓存，由于内核缓存是全局的，因此，这个方法不是针对WebView，而是针对整个应用程序。

(2) public void clearFormData()

作用：仅仅清除自动完成填充的表单数据，并不会清除WebView存储到本地的数据。

(3) public void clearHistory()

作用：清除当前WebView访问的历史记录，只会清除当前页之前的历史记录。为了解决clearHistory()的失效问题，在调用loadUrl()方法后，一般采用如下方法：

```
mWebView.postDelayed(new Runnable() {
    @Override
```

```
    public void run() {
        mWebView.clearHistory();
    }
}, 1000);
```

(4) public void clearMatches()

作用：清除网页查找的高亮匹配字符。

6. WebView 的状态

(1) public void onResume()

作用：激活 WebView 为活跃状态，能正常执行网页的响应。

(2) public void onPause()

作用：当页面失去焦点被切换到后台不可见状态时，需要执行 onPause()方法，该方法会通知内核暂停所有的动作，比如 DOM 的解析、plugin 的执行、JavaScript 的执行。并且可以减少不必要的 CPU 和网络开销，可以达到省电、省流量、省资源的效果。

(3) public void destroy()

作用：这个方法必须在 WebView 从 view tree 中删除之后才能被执行，这个方法会通知 native 释放 WebView 占用的所有资源。

7. WebView 事件回调监听

(1) public void setWebChromeClient(WebChromeClient client)

作用：主要通知客户端 App 加载当前网页的对话框、网站图标、网站标题、加载进度等事件，通知客户端处理这些响应的事件。例如：

```
mWebView.setWebChromeClient(new WebChromeClient() {
    public void onProgressChanged(WebView view, int progress) {
        setProgress(progress * 100);
        if(progress == 100){
            imageView1.setVisibility(View.GONE);
            tv1.setVisibility(View.GONE);
            pb1.setVisibility(View.GONE);
            fy1.setVisibility(View.GONE);
        }
    }
});
```

(2) public void setWebViewClient(WebViewClient client)

作用：主要通知客户端 App 加载当前网页时的各种状态，如 onPageStart、onPageFinish、onReceiveError 等事件。例如：

```
mWebView.setWebViewClient(new WebViewClient() {
    public void onReceivedError(WebView view, int errorCode,
            String description, String failingUrl) {
        //Handle the error
        Toast.makeText(getApplicationContext(), "网络连接失败,请连接网络。",
```

```
                Toast.LENGTH_SHORT).show();
    }

    public boolean shouldOverrideUrlLoading(WebView view, String url) {
        view.loadUrl(url); return true;
    }
});
```

6.2.3 页面导航

当用户在WebView中点击一个链接时,对于Android来说默认的动作是启动一个应用程序捕捉它。通常,默认的网页浏览器会打开并且载入目标URL。然而,可以在WebView中重写该动作,通过调用setWebViewClient()方法把该链接在WebView中打开。还可以让用户自己处理历史的回退和前进功能。

如果希望点击页面中的链接继续在当前浏览器中响应,而不是启动Android系统中的浏览器响应该链接,必须覆盖WebView的WebViewClient对象。例如:

```
webView.setWebViewClient(new WebViewClient(){
    @Override
    public boolean shouldOverrideUrlLoading(WebView view, String url) {
        view.loadUrl(url);
        return true;
    }
});
```

如果要在点击链接并载入网页的时候做更多的操作,可以创建自己的WebViewClient并重写shouldOverrideUrlLoading()方法。例如:

```
private class MyWebViewClient extends WebViewClient {
    @Override
    public boolean shouldOverrideUrlLoading(WebView view, String url) {
        //处理自己的页面
        if (Uri.parse(url).getHost().equals("www.google.com")) {
            return false;
        }
        //处理其他网页,则启动另外的Activity来处理该URL
        Intent intent = new Intent(Intent.ACTION_VIEW, Uri.parse(url));
        startActivity(intent);
        return true;
    }
}
```

然后给WebView创建一个WebViewClient的实例。

```
webView.setWebViewClient(new MyWebViewClient());
```

这时,当用户点击一个链接,系统就会调用 shouldOverrideUrlLoading(),它会辨别 URL 主机是否要匹配指定的域名。如果它不能匹配,则该方法返回 false,一个 Intent 会创建并启动默认的 Activity 来处理该 URL(它一般解析为用户的默认浏览器)。

用 WebView 点击链接访问多个网页后,默认情况下,点击系统返回键,整个浏览器会调用 finish()而结束自身,WebView 不会回退到上一个页面。为了让 WebView 支持回退功能,需要覆盖 Activity 类的 onKeyDown()方法,可以使用 goBack()方法和 goForward()方法操纵回退和向前功能。例如:

```
@Override
public boolean onKeyDown(int keyCode, KeyEvent event) {
    if ((keyCode == KeyEvent.KEYCODE_BACK) && myWebView.canGoBack() {
        webView.goBack();
        return true;
    }
    return super.onKeyDown(keyCode, event);
}
```

如果有网页历史,则 canGoBack()方法返回 True。同样,可以使用 canGoForward()来检查是否有"前进"历史。如果不执行这个检查,当用户到达浏览历史尽头的时候,goBack()和 goForward()什么都不做。

6.2.4 WebSettings 与缓存处理

android.webkit.WebSettings 是 WebView 组件的一个辅助类,用于管理 WebView 的设置状态。该对象可以通过 WebView.getSettings()方法获得。

WebSettings 提供的核心方法包括以下几种。

- setAllowFileAccess(boolean allow):启用或禁止 WebView 访问文件数据。
- setBlockNetworkImage(boolean allow):确定是否显示网络图像。
- setLoadsImagesAutomatically(boolean flag):设置是否支持自动加载图像。
- setBuiltInZoomControls(boolean enabled):设置是否支持缩放。
- setCacheMode(int mode):设置缓冲的模式。
- setDefaultFontSize(int size):设置默认字体的大小。
- setDefaultTextEncodingName(String encoding):设置在解码时使用的默认编码。
- setFixedFontFamily(String font):设置固定使用的字体。
- setJavaSciptEnabled(boolean flag):设置是否支持 Javascript。
- setJavaScriptCanOpenWindowsAutomatically(boolean flag):设置是否支持通过 JS 打开新窗口。
- setLayoutAlgorithm(WebSettings.LayoutAlgorithm l):设置布局方式。
- setLightTouchEnabled(boolean enabled):设置用鼠标激活被选项。
- setSupportZoom(boolean support):设置是否支持变焦。
- setPluginsEnabled(boolean flag):设置是否支持插件。
- supportMultipleWindows():多窗口支持。

例如，打开页面时，自适应屏幕的代码如下：

```
WebSettings webSettings = mWebView.getSettings();
webSettings.setUseWideViewPort(true);   //设置此属性,可任意比例缩放
webSettings.setLoadWithOverviewMode(true);
```

下面的代码使得页面支持缩放：

```
WebSettings webSettings =   mWebView.getSettings();
webSettings.setJavaScriptEnabled(true);
webSettings.setBuiltInZoomControls(true);
webSettings.setSupportZoom(true);
```

下面介绍一下关于浏览网页时缓存的处理。

① 打开及关闭缓存。

优先使用缓存的代码如下：

```
webView.getSettings().setCacheMode(WebSettings.LOAD_CACHE_ELSE_NETWORK);
```

不使用缓存的代码如下：

```
webView.getSettings().setCacheMode(WebSettings.LOAD_NO_CACHE);
```

WebSettings.CacheMode 总共有以下 4 个选项。

- LOAD_DEFAULT：这是默认加载方式，使用这种方式会实现快速前进或后退。在同一个标签中打开使用几个网页后，关闭网络时，可以通过前进、后退来切换已经访问过的数据。同时新建网页时需要使用网络。
- LOAD_NO_CACHE 和 LOAD_NORMAL：这两种方式不使用缓存。如果没有网络，即使以前打开过此网页，也不会使用以前的网页。
- LOAD_CACHE_ELSE_NETWORK：这个方式总是会从缓存中加载，除非缓存中的网页过期，出现的问题就是打开动态网页时不能时时更新，并且会出现上次打开过的状态，除非清除了缓存。
- LOAD_CACHE_ONLY：这个方式只会使用缓存中的数据，不会使用网络。

影响缓存模式的两个 http 头是 If-None-Match 和 If-Modified-Since，遇到这两个 http 头，浏览器会把缓存模式改为 LOAD_NO_CACHE。

② 在退出应用程序时删除缓存的代码如下：

```
File file = CacheManager.getCacheFileBaseDir();
if (file != null && file.exists() && file.isDirectory()) {
   for (File item : file.listFiles()) {
        item.delete();
   }
   file.delete();
```

```
}
context.deleteDatabase("webview.db");
context.deleteDatabase("webviewCache.db");
```

③ 删除保存在手机上的缓存代码如下：

```
private int clearCacheFolder(File dir, long numDays) {
    int deletedFiles = 0;
    if (dir!= null && dir.isDirectory()) {
        try {
            for (File child: dir.listFiles()) {
                if (child.isDirectory()) {
                    deletedFiles += clearCacheFolder(child, numDays);
                }

                if (child.lastModified() < numDays) {
                    if (child.delete()) {
                        deletedFiles++;
                    }
                }
            }
        } catch(Exception e) {
            e.printStackTrace();
        }
    }
    return deletedFiles;
}
```

6.2.5 WebChromeClient 和 WebViewClient

与 WebView 相关的辅助对象，除了 WebSettings 以外，还有 WebChromeClient 和 WebViewClient。

1. WebChromeClient

WebChromeClient 主要用来辅助 WebView 处理 Javascript 的对话框、网站图标、网站标题以及网页加载进度等。

同样地，可以通过 WebView 的 setWebChromeClient()方法为 WebView 对象指定一个 WebChromeClient。

在 WebChromeClient 中，当网页的加载进度发生变化时，onProgressChanged(WebView view, int newProgress)方法会被调用；当网页的图标发生改变时，onReceivedIcon(WebView view, Bitmap icon)方法会被调用；当网页的标题发生改变时，onReceivedTitle(WebView view, String title)方法会被调用。利用这些方法，便可以很容易地获得网页的加载进度、网页的图标和标题等信息，例如：

```
MyWebChromeClient myWebChromeClient = new MyWebChromeClient();
mWebView.setWebChromeClient(myWebChromeClient);
```

```
private class MyWebChromeClient extends WebChromeClient {

    //获得网页的加载进度,显示在右上角的 TextView 控件中
    public void onProgressChanged(WebView view, int newProgress) {
        if(newProgress < 100) {
            String progress = newProgress + "%";
            mTextView_progress.setText(progress);
        } else {
            mTextView_progress.setText(" ");
        }
    }

    //获得网页的标题,作为应用程序的标题进行显示
    public void onReceivedTitle(WebView view, String title) {
        MainActivity.this.setTitle(title);
    }

}
```

2. WebViewClient

WebViewClient 类定义了一系列事件方法，如果 Android 应用程序设置了 WebVieClient 派生对象，则在网页载入、资源载入、页面访问错误等情况发生时，该派生对象的相应方法会被调用。

WebViewClient 常用方法包括以下几种。

- doUpdateVisitedHistory(WebView view, String url, boolean isReload)：更新历史记录。
- onFormResubmission(WebView view, Message dontResend, Message resend)：应用程序重新请求网页数据。
- onLoadResource(WebView view, String url)：该方法在加载页面资源时会被调用，每一个资源(比如图像)的加载都会调用一次。
- onPageStarted(WebView view, String url, Bitmap favicon)：这个事件就是开始载入页面时调用的，通常可以在这里设定一个 loading 页面，告诉用户程序在等待网络响应。
- onPageFinished(WebView view, String url)：在页面加载结束时调用该方法。一个页面载入完成后，可以关闭 loading 条并切换程序的动作。
- onReceivedError(WebView view, int errorCode, String description, String failingUrl)：报告错误信息。
- onReceivedHttpAuthRequest(WebView view, HttpAuthHandler handler, String host,String realm)：获取返回信息授权请求。
- onReceivedSslError(WebView view, SslErrorHandler handler, SslError error)：重写此方法可以让 WebView 处理 https 请求。
- onScaleChanged(WebView view, float oldScale, float newScale)：WebView 发生

改变时调用该方法。

- onUnhandledKeyEvent(WebView view, KeyEvent event)：Key 事件未被加载时调用该方法。
- shouldOverrideKeyEvent(WebView view, KeyEvent event)：重写此方法才能够处理在浏览器中的按键事件。
- shouldOverrideUrlLoading(WebView view, String url)：在点击请求链接时才会调用,重写此方法则返回 true,表明无论是点击网页里面的链接还是在当前的 WebView 里跳转,都会跳到浏览器中。

代码举例如下：

```
WebViewClient wvc = new WebViewClient() {

    @Override
    public boolean shouldOverrideUrlLoading(WebView view, String url) {
        Toast.makeText(getApplicationContext(),
                "WebViewClient.shouldOverrideUrlLoading",
                Toast.LENGTH_SHORT)
                .show();
        //使用自己的 WebView 组件来响应 Url 加载事件,而不是使用默认浏览器加载页面
        wv.loadUrl(url);
        return true;
    }

    @Override
    public void onPageStarted(WebView view, String url, Bitmap favicon) {
        Toast.makeText(getApplicationContext(),
                "WebViewClient.onPageStarted",
                Toast.LENGTH_SHORT).show();
        super.onPageStarted(view, url, favicon);
    }

    @Override
    public void onPageFinished(WebView view, String url) {
        Toast.makeText(getApplicationContext(),
                "WebViewClient.onPageFinished",
                Toast.LENGTH_SHORT).show();
        super.onPageFinished(view, url);
    }

    @Override
    public void onLoadResource(WebView view, String url) {
        Toast.makeText(getApplicationContext(),
                "WebViewClient.onLoadResource",
                Toast.LENGTH_SHORT).show();
        super.onLoadResource(view, url);
    }

};
```

6.3 使用 HTML5 开发 Web App

HTML5 针对移动 Web 应用程序引入了大量新特性，其中包括离线缓存技术、音频视频自由嵌入、地理位置获取、Canvas 绘图、丰富的表单元素及交互方式等，另外，结合 Jquery Mobile 等框架提供的可视化特性，使得应用程序给用户带来强烈的视觉冲击。

6.3.1 使用 JavaScript 访问 Android

使用 WebView 进行移动应用开发的目的，不仅仅是作为查看网站的手段。相反，应用程序嵌入的网页应该针对环境而进行设计，甚至可以定义 Android 应用程序和网页之间的接口，这一接口允许 JavaScript 调用 Android 应用程序 API——向基于 Web 的应用程序提供支持。

如果载入的网页实现了 JavaScript，必须在 WebView 中开启 JavaScript 功能，并在应用代码和 JavaScript 代码之间创建接口。步骤如下：

1. 开启 JavaScript

在 WebView 中 JavaScript 默认是不开启的。可以通过 WebSettings(使用 getSettings()方法获得 WebSettings)设置 WebView 对 JavaScript 的开启，然后使用 setJavaScriptEnabled()来开启它。例如：

```
WebView mWebView = (WebView) findViewById(R.id.webview);
WebSettings webSettings = mWebView.getSettings();
webSettings.setJavaScriptEnabled(true);
```

在 WebSettings 中已经实现了很多有用的功能。例如，如果使用 WebView 开发一个网络应用，可以使用 setUserAgentString()方法定义一个自定义用户代理字符串，后面可以查询网页中的自定义用户代理，以便辨别当前请求网页的是不是 Android 应用。

2. 绑定 JavaScript 到 Android 中

当使用 WebView 创建网络应用的时候，可以在 JavaScript 和 Android 源代码之间创建一个接口，例如，当点击页面中的电话号码时，让 JavaScript 代码调用 Android 中定义的方法来实现拨号功能。例如：

```
public class JavaScriptInterface {
    Context mContext;

    JavaScriptInterface(Context c) {
        mContext = c;
    }

    public void startPhone(String num){
        Intent intent = new Intent();
        intent.setAction(Intent.ACTION_CALL);
```

```
        intent.setData(Uri.parse("tel:" + num));
        startActivity(intent);
    }

}
```

在 WebView 中使用 addJavascriptInterface(Object obj,String interfaceName)方法将一个 Java 对象绑定到一个 Javascript 对象中,Javascript 对象名就是 interfaceName,作用域是全局的。例如:

```
webView.addJavascriptInterface(new JavaScriptInterface(this), "Android");
```

这段代码创建了一个名称为"Android"的 JavaScript 接口并运行在 WebView 当中。Web 应用会接入 JavaScriptInterface 类。例如,下面的代码会在用户点击电话号码链接时启动 Android 自带的拨号程序。

```
<!DOCTYPE html PUBLIC "-//W3C//DTD HTML 4.01 Transitional//EN"
        "http://www.w3.org/TR/html4/loose.dtd">
<html>
    <head>
        <meta http-equiv = "Content-Type" content = "text/html; charset = UTF-8">
            <title>Javascript Demo</title>
        </head>
    <body>
            <table align = "center">
                <tr><td>部门</td><td>电话</td></tr>
                <tr><td>移动</td><td>
                    <a  href = "javascript:Android.startPhone(10086)">10086</a>
                </td></tr>
            </table>
    </body>
</html>
```

不需要初始化 JavaScript 中的 Android 接口。WebView 会自动让它可以应用于网页当中。

注意:在 JavaScript 中绑定的对象会运行在另一个线程,它和创建它的线程是不同的线程。使用 addJavaScriptInterface()允许 JavaScript 来控制 Android 应用。

6.3.2 使用 CSS 适配 UI

当 Android 浏览器加载一个网页时,默认是以"概览模式"加载这个页面,"概览模式"是指提供缩小至这个页面的远景的视图。Android 浏览器和 WebView 会通过缩放网页补偿屏幕密度的变化,使得所有的设备在显示此网页时看上去和中等密度下显示的大小相同。

Android 浏览器和 WebView 支持 CSS 媒介类型,可以使用-webkit-device-pixel-ratio 的 CSS 媒介类型为特定的屏幕密度创建一个样式。允许使用的值包括 0.75、1 或 1.5,分别表示低、中和高密度屏幕。

例如，可以分别为每个密度建立样式表。

```
<link rel="stylesheet"
   media="screen and (-webkit-device-pixel-ratio: 1.5)"
   href="hdpi.css" />
<link rel="stylesheet"
   media="screen and (-webkit-device-pixel-ratio: 1.0)"
   href="mdpi.css" />
<link rel="stylesheet"
   media="screen and (-webkit-device-pixel-ratio: 0.75)"
   href="ldpi.css" />
```

或者在一个样式表中指定不同的样式，例如：

```
#header {
   background:url(medium-density-image.png);
}

@media screen and (-webkit-device-pixel-ratio: 1.5) {
   /* CSS for high-density screens */
   #header {
       background:url(high-density-image.png);
   }
}

@media screen and (-webkit-device-pixel-ratio: 0.75) {
   /* CSS for low-density screens */
   #header {
       background:url(low-density-image.png);
   }
}
```

为了提供完全自定义的样式，使页面是为每个支持的密度定制的，应该设置 viewport 属性，使 viewport 的宽度和密度和设备相匹配。通过这种方式，Android 浏览器和 WebView 不对网页进行缩放，并且 viewport 宽度完全匹配于屏幕宽度。但是通过添加一些使用-webkit-device-pixel-ratio 媒介类型的自定义 CSS，可以使用不同的样式。

6.3.3 jQuery Mobile 框架

jQuery Mobile 是一个针对触摸体验的 Web UI 开发框架，实现开发跨智能电话和平板电脑工作的移动 Web 应用程序。jQueryMobile 框架构建于 jQuery 内核之上，提供了构建完整移动 Web 应用程序和网站所需的所有 UI 组件，包括页面、对话框、工具栏、不同类型的列表视图，各种表单元素和按钮等。

jQuery Mobile 基本特性包括以下几种。

1. 一般简单性和灵活性

该框架易于使用，可以使用标记驱动开发页面，无须或仅需很少的 JavaScript；使用高级 JavaScript 和事件；使用一个 HTML 文件和多个嵌入页面；将应用程序分解成多个页面。

2. 逐步强化和全面兼容

尽管 jQuery Mobile 利用最新的 HTML5、CSS3 和 JavaScript，但并非所有移动设备都提供这样的支持。jQuery Mobile 的理念是同时支持高端和低端设备，比如那些不支持 JavaScript 的设备，尽量提供最好的体验。

3. 支持触摸屏输入和其他输入方法

jQuery Mobile 为不同输入方法和事件提供支持：触摸屏、鼠标和基于光标焦点的用户输入。

4. 可访问性

jQuery Mobile 在设计时考虑了访问能力，它支持 Accessible Rich Internet Applications (WAI-ARIA)，以帮助使用辅助技术的残障人士访问 Web 页面。

5. 轻量级和模块化

该框架属于轻量级，拥有一个大小为 24KB 的 JavaScript 库、7KB 的 CSS 以及一些图标。

6. 主题

该框架还提供一个主题系统，允许定义自己的应用程序样式。

下面以一个图像浏览应用来介绍基于 HTML5＋jQuery Mobile 开发 Web App 的方法。

Activity 实例类核心代码如下：

```
public class HTML5Activity extends Activity {

    private WebView webView;

    @Override
    protected void onCreate(Bundle savedInstanceState) {
        super.onCreate(savedInstanceState);

        setContentView(R.layout.activity_html5);
        webView = (WebView) findViewById(R.id.fullscreen_content);
        webView.setWebViewClient(new MyWebViewClient());
        webView.loadUrl("file:///android_asset/www/index.html");
        webView.setWebChromeClient(new WebChromeClient());
        WebSettings webSettings = webView.getSettings();
        webSettings.setJavaScriptEnabled(true);
        webSettings.setDomStorageEnabled(true);
        webSettings.setDatabasePath("/data/data/" + this.getPackageName()
                + "/databases/");

    }

    @Override
    public void onBackPressed() {
```

```
24          if (webView.canGoBack())
25              webView.goBack();
26          else
27              super.onBackPressed();
28      }
29
30      @Override
31      protected void onPostCreate(Bundle savedInstanceState) {
32          super.onPostCreate(savedInstanceState);
33      }
34  }
```

其中，第 11 行的 MyWebViewClient()代码如下：

```
public class MyWebViewClient extends WebViewClient {
    @Override
    public boolean shouldOverrideUrlLoading(WebView view, String url) {
        view.loadUrl(url);
        return true;
    }
}
```

在 Intel XDK[①] 中，选择 Start with App Designer 来创建 HTML5 项目，项目框架选择 jQuery Mobile，如图 6-3 所示。

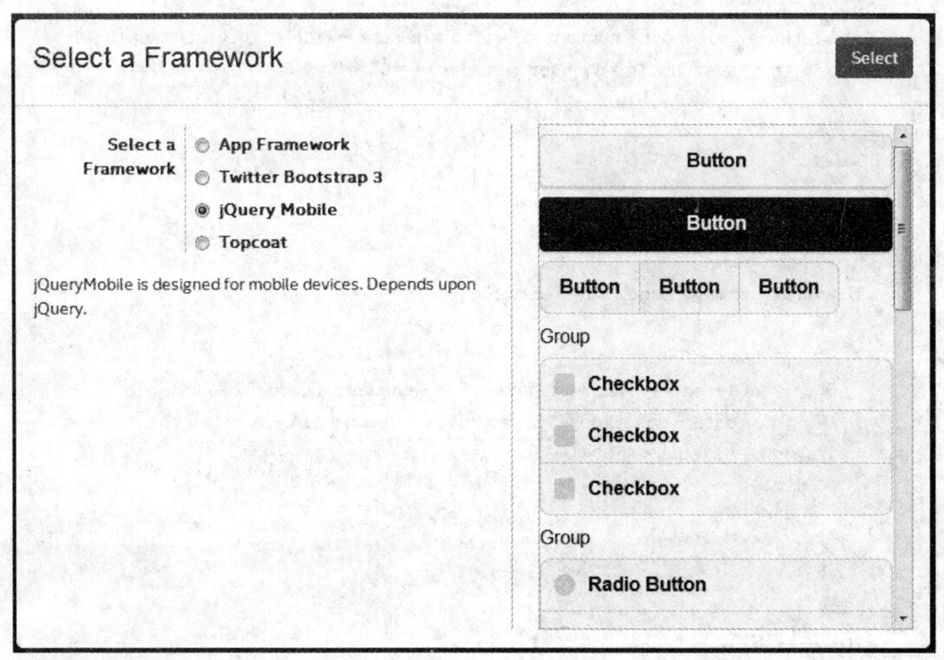

图 6-3　Web App 设计效果

① 开发环境下载地址为 https://software.intel.com/en-us/html5/tools。

实现 index.html 主页面的核心代码如下：

```
<!DOCTYPE html>
<html>
<head>
    <meta name = "Copyright" content = "&copy; 2012, Intel Corporation.
              All rights reserved." />

    <script type = "text/javascript"
            src = "vendor/jquery/jquery-1.8.0.js"></script>
    <script type = "text/javascript" src = "app/config.js"></script>
    <script type = "text/javascript"
            src = "vendor/jquery.mobile/jquery.mobile-1.1.1.js"></script>
    <script type = "text/javascript" src = "app/tabbedimages.js"></script>
    <script type = "text/javascript" src = "app/optionsWidget.js"></script>
    <script src = "corodova.js"></script>
    <script src = "intelxdk.js"></script>
    <script src = "xhr.js"></script>

    <script type = "text/javascript">
        var onDeviceReady = function(){
            if( window.Cordova && navigator.splashscreen ) {
                navigator.splashscreen.hide();
            }
        };
        document.addEventListener("deviceready", onDeviceReady, false);
        </script>
    <meta name = "viewport" content = "width = device-width, initial-scale = 1,
            maximum-scale = 1, user-scalable = 0">

</head>

<body>

    <!-- home page -->
    <div data-role = "page" id = "birds">

        <!-- header -->
        <div data-role = "header" data-id = "tabnav-header"
                data-position = "fixed" data-theme = "b">
            <h1>Albums</h1>
        </div>
        ...
        </div>

    <!-- page 2 -->
    <div data-role = "page" id = "flowers">
    ...
    </div>

```

```
    <!-- page 3 -->
    <div data-role="page" id="animals">
    ...
    </div>
    ...

    <script type="text/javascript">
        $("#birds").die("pageinit");
        $("#flowers").die("pageinit");
        $("#animals").die("pageinit");
    </script>

</body>

</html>
```

CSS 样式核心代码如下：

```
body {
  /*   background: #232323 !important; */
background: #b4e391; /* Old browsers */
background: -moz-linear-gradient(top,  #b4e391 0%, #61c419 50%,
     #b4e391 100%);
background: -webkit-gradient(linear, left top, left bottom,
     color-stop(0%, #b4e391), color-stop(50%, #61c419),
     color-stop(100%, #b4e391));
...
}

/* ensure background takes up entire screen */
html, body {
     width: 100%;
     height: 100%;
     overflow: hidden;
     margin: 0px !important;
     padding: 0px !important;
}

.content_div{
      position: fixed;
      width: 100%;
      padding: 0px !important;
      margin: 0px !important;

}

.mainimage_div{
    position: fixed;
    padding: 0px !important;
```

```
        margin: 0px !important;
        width: 100%;
        top: 36px;

    /*  height: 80%;
        width: 95%; */
}

.thumbnails_div{
        position: fixed;
        padding: 0px !important;
        margin: 0px !important;
        width: 100%;
/* padding-bottom: 36px; */


}

.mainimage{
        max-width: 100%;
        max-height: 100%;
    /*  top-margin: -30px; */
    padding: 0px !important;
    margin-top: 0px !important;
    margin-bottom: 0px !important;
        margin-left: auto;
        margin-right: auto;


        display: block;
}

.thumbnail:hover{ opacity: 0.7; }
.thumbnail:active { opacity: 0.7;}

.thumbnail {
        max-height: 100%;
        max-width: 15%;
    /*  min-width: 46px; */
        width: 12%;

}
.highlight{
        opacity: 0.5;
}

/* override default jQuery Mobile page styling */
.ui-page {
        /* transparent page (to go over linen background) with white font */
```

```
        background: none;
        background-image: none;
        color: white;

        /* hacks to get smoother page transition, esp on older Androids */
        -webkit-backface-visibility: hidden;
        -moz-backface-visibility: hidden;
        -ms-backface-visibility: hidden;
        -o-backface-visibility: hidden;
        backface-visibility: hidden;
}

/* custom styling of options menu (position & width) */
.optionsMenu {
    width: 120px;
    position: fixed;
    right: 1px;
}

.optionsMenu .ui-field-contain {
    margin: 0;
    padding: 0;
}

.optionsMenu .ui-controlgroup-controls {
    width: 100%;
}

/* no rounded corners for options menu */
.optionsMenu .ui-corner-top, .ui-corner-bottom {
    -webkit-border-radius: 0px !important;
    -moz-border-radius: 0px !important;
    -ms-border-radius: 0px !important;
    -o-border-radius: 0px !important;
    border-radius: 0px !important;
}
```

config.js 核心代码如下：

```
$(document).bind('mobileinit', function () {
    $.mobile.defaultPageTransition = 'none';
});

/* keep header and footer visible at all times */
$(document).on('pageinit', ':jqmData(role=page)', function() {
    $(this).find(':jqmData(role=header)').fixedtoolbar( {
            tapToggle: false } );
    $(this).find(':jqmData(role=footer)').fixedtoolbar( {
            tapToggle: false } );
});
```

将 XDK 中的项目复制到 Android 项目的 assets 文件夹下，并运行项目。设计效果如图 6-4 所示。

图 6-4　Web App 设计效果

6.4　习　题

1. 编程实现基于 WebView 的地图设计。
2. 编程实现基于 HTML5 技术在 WebView 中播放视频。

第 7 章　开放接口编程

作为 Internet 异构环境下的互操作技术，Web 服务被广泛应用，并衍生出基于开放接口的网络编程技术。

7.1　Web 服务编程

Web 服务（WebService 或 Web Service）是可供外部使用的分布式应用程序组件，通过 Web 服务实现不同系统之间的相互调用，从而实现系统集成的平台无关性、语言无关性。

7.1.1　Web 服务概述

Web 服务是一种服务导向架构的技术，通过标准的 Web 协议提供服务，目的是保证不同平台的应用服务可以互操作。

1. Web 服务协议

根据 W3C 的定义，Web 服务应当是一个软件系统，用以支持网络间不同机器的互动操作。Web 服务通常是由许多应用程序接口（API）所组成的，它们通过网络执行客户所提交服务的请求，使调用者能够用编程的方式通过 Web 调用来开发应用程序。

Web 服务也遵循 Web 通信协议，如 HTTP、TCP/IP、SMTP 等。不同的系统需要遵守同一种协议来发送和接收 XML 数据（由于 Web 服务要最终实现跨平台、跨语言之间的互相通信和数据共享，数据的传输必须以一定的格式和标准进行，这种标准就是 XML），达到通信的目的，这个协议是 SOAP（Simple Object Access Protocl，简单对象访问协议）。

各类 Web 服务框架的本质就是一个大的 Servlet，当远程客户端通过 HTTP 协议发送来 SOAP 格式的请求数据时，它分析这个数据，就知道要调用服务端的哪个 Java 类以及哪个方法。当确定了服务端 Java 类以及方法后，服务就去查找或者创建这个对象，并调用其方法，再把方法返回的结果包装成 SOAP 格式的数据，通过 HTTP 协议将消息发给客户端。

2. Web 服务体系结构

Web 服务的体系是一种面向服务的体系结构。包括三种角色（见图 7-1）。

(1) 服务提供商

发布自己的服务，并且对服务请求进行响应。

开发完 Web 服务后，需要一种发布手段让自己的 Web 服务被网络上任何一个地点的人或者系

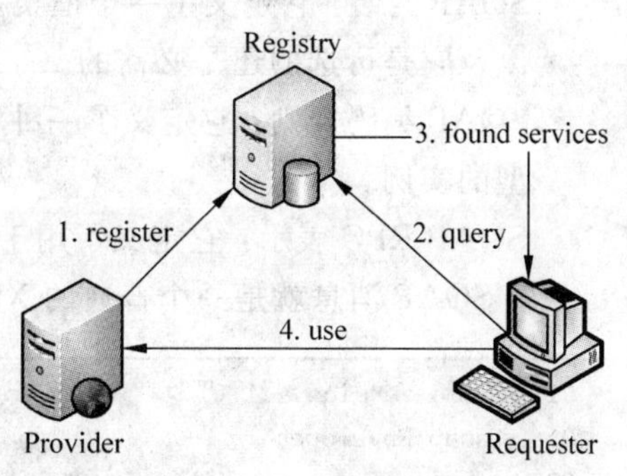

图 7-1　Web 服务体系结构

统得知从而调用它,这个手段是 UDDI(通用描述、发现与集成)。服务提供者要在 UDDI 注册中心注册。

(2) 服务代理商

注册已经发布的 Web 服务,对其进行分类,并提供搜索服务。

UDDI 注册中心扮演了服务"代理者"的角色,它通过 WSDL(Web Service Description Language,Web 服务描述语言)来向服务请求者展示已经注册的 Web 服务。

(3) 服务请求者

利用服务代理商查找所需的服务,然后使用该服务。

也就是在 UDDI 中心,通过 WSDL 查找已经发布的 Web 服务,以便自己使用。

3. Web 服务编程模型

为 Web 服务开发者提供了多个编程模型。这些模型分为两类。

(1) 基于 REST

表述性状态传递(Representational State Transfer,REST)是一种创建并与 Web 服务通信的新方法。在 REST 中,资源具有 URI 并通过 HTTP 头操作进行处理。每个文档和每个流程都通过唯一的 URI 建模为 Web 资源。这些 Web 资源由可以在 HTTP 头中指定的操作处理。HTTP 是 REST 中的协议。仅有四种方法可用:GET、PUT、POST 和 DELETE。可以对请求标记书签并可以缓存响应。网络管理员只需通过查看 HTTP 头,就可以方便地跟踪 REST 风格的服务。

(2) 基于 SOAP/WSDL

在基于 SOAP 的 Web 服务中,Java 实用程序基于 Web 服务中的 Java 代码创建 WSDL 文件。WSDL 可以在网络上公开。对使用 Web 服务感兴趣的各方可以基于 WSDL 创建 Java 客户端。消息以 SOAP 格式进行交换。可以传入 SOAP 的操作范围比 REST 中提供的要宽得多,特别是在安全性方面。

7.1.2 核心技术

1. SOAP

SOAP 指简单对象访问协议(Simple Object Access Protocol),它是一种基于 XML 的可扩展消息信封格式,用于网络上不同平台、不同语言的应用程序间的通信。

SOAP 包括三个部分。

- SOAP 封装:它定义了一个框架,该框架描述了消息中的内容是什么,谁应当处理它以及它是可选的还是必需的。
- SOAP 编码规则:它定义了一种序列化的机制,用于交换应用程序所定义的数据类型的实例。
- SOAP RPC 表示:它定义了用于表示远程过程调用和应答的协定。

一条 SOAP 消息就是一个普通的 XML 文档,基本结构如下:

```
<? xml version = "1.0"?>
<soap:Envelope
        xmlns:soap = "http://www.w3.org/2001/12/soap - envelope"
```

```
        soap:encodingStyle = "http://www.w3.org/2001/12/soap - encoding">
    < soap:Header >
        ...
    </soap:Header >
    < soap:Body >
        ...
        < soap:Fault >
            ...
        </soap:Fault >
    </soap:Body >
</soap:Envelope >
```

SOAP 消息必须用 XML 来编码，且必须使用 Envelope 命名空间和 Encoding 命名空间。SOAP 消息不能包含 DTD 引用，也不能包含 XML 处理指令。

SOP 中各元素的含义如下。

- Envelope 元素：标识 XML 文档的一条 SOAP 消息。
- Header 元素：包含头部信息的 XML 标签。
- Body 元素：包含所有的调用和响应的主体信息的标签。
- Fault 元素：错误信息标签。

下面的例子中，一个 GetStockPrice 请求被发送到了服务器。此请求有一个 StockName 参数，而在响应中则会返回一个 Price 参数。

- SOAP 请求。

```
POST /InStock HTTP/1.1
Host: www.jsoso.net
Content - Type: application/soap + xml; charset = utf - 8
Content - Length: × × ×

<? xml version = "1.0"?>
< soap:Envelope
        xmlns:soap = "http://www.w3.org/2001/12/soap - envelope"
        soap:encodingStyle = "http://www.w3.org/2001/12/soap - encoding">
    < soap:Body xmlns:m = "http://www.jsoso.net/stock">
        < m:GetStockPrice >
            < m:StockName > IBM </m:StockName >
        </m:GetStockPrice >
    </soap:Body >
</soap:Envelope >
```

- SOAP 响应。

```
HTTP/1.1 200 OK
Content - Type: application/soap + xml; charset = utf - 8
Content - Length: × × ×

```

```
<? xml version = "1.0"?>
< soap:Envelope
       xmlns:soap = "http://www.w3.org/2001/12/soap - envelope"
       soap:encodingStyle = "http://www.w3.org/2001/12/soap - encoding">
   < soap:Body xmlns:m = "http://www.jsoso.net/stock">
       < m:GetStockPriceResponse >
           < m:Price > 34.5 </m:Price >
       </m:GetStockPriceResponse >
   </soap:Body >
</soap:Envelope >
```

2. WSDL

WSDL(Web Service Description Language,Web 服务描述语言)是一种 XML 格式文档,用以描述服务端口访问方式和使用协议的细节。通常用来辅助生成服务器和客户端代码及配置信息。WSDL 文档包括访问和使用 Web 服务所必需的信息,定义该 Web 服务的位置、功能以及如何通信等描述信息。

一个 WSDL 文档在定义网络服务时使用如下的元素。

- definitions：WSDL 文档的根元素,该元素的属性指明了 WSDL 文档的名称、文档的目标名字空间,以及 WSDL 文档应用的名字空间的速记定义。
- types：使用某种方式(如 XSD)的数据类型定义集合,形成服务所用消息的构建块。
- message：用于通信的抽象数据类型定义,即参数。
- operation：一个服务包含的操作的描述,当操作被调用时,操作被定义为两个低端之间的消息传递。
- portType：描述服务逻辑接口的 operation 元素的集合。
- binding：特定端口类型所使用的具体协议和数据格式规范。一个 binding 元素定义如何将一个抽象消息映射到一个具体数据格式。该元素指明诸如参数顺序、返回值等信息。
- port：绑定和网络地址关联的单一接入点,这个元素将所有抽象定义聚集在一起。
- service：元素包含一系列的 port 子元素,port 子元素将会把绑定机制、服务访问协议和端点地址结合在一起。

一般来说,只要调用者能够获取 Web 服务对应的 WSDL,就可以从中了解它所提供的服务以及如何调用 Web 服务。因为一份 WSDL 文件清晰地定义了以下三个方面的内容。

- WHAT 部分：用于定义 Web 服务所提供的操作(或方法),也就是 Web 服务能做些什么。由 WSDL 中的 types、message 和 portType 元素定义。
- HOW 部分：用于定义如何访问 Web 服务,包括数据格式详情和访问 Web 服务操作的必要协议。也就是定义了如何访问 Web 服务。由 binding 元素定义。
- WHERE 部分：用于定义 Web 服务位于何处,如何使用特定协议决定的网络地址(如 URL)指定。该部分由 service 元素定义,可在 WSDL 文档的最后部分看到 service 元素。

以下是中国气象局 Web 服务的 WSDL 文档框架：

```
<wsdl:definitions xmlns:soap = "http://schemas.xmlsoap.org/wsdl/soap/"
        xmlns:tm = "http://microsoft.com/wsdl/mime/textMatching/"
        xmlns:soapenc = "http://schemas.xmlsoap.org/soap/encoding/"
        xmlns:mime = "http://schemas.xmlsoap.org/wsdl/mime/"
        xmlns:tns = "http://WebXml.com.cn/"
        xmlns:s = "http://www.w3.org/2001/XMLSchema"
        xmlns:soap12 = "http://schemas.xmlsoap.org/wsdl/soap12/"
        xmlns:http = "http://schemas.xmlsoap.org/wsdl/http/"
        xmlns:wsdl = "http://schemas.xmlsoap.org/wsdl/"
        targetNamespace = "http://WebXml.com.cn/">
   ...
   <wsdl:types>
        <s:schema elementFormDefault = "qualified"
                targetNamespace = "http://WebXml.com.cn/">
            <s:element name = "getSupportCity">
                <s:complexType>
                    <s:sequence>
                        <s:element minOccurs = "0" maxOccurs = "1"
                                name = "byProvinceName" type = "s:string"/>
                    </s:sequence>
                </s:complexType>
            </s:element>
   </wsdl:types>
   <wsdl:service name = "WeatherWebService">
        <wsdl:port name = "WeatherWebServiceHttpPost"
                binding = "tns:WeatherWebServiceHttpPost">
            <http:address location = "http://www.webxml.com.cn/
                    WebServices/WeatherWebService.asmx"/>
        </wsdl:port>
   </wsdl:service>
</wsdl:definitions>
```

其中，types 元素使用 XML 模式语言声明在 WSDL 文档中的其他位置使用的复杂数据类型与元素；import 元素类似于 XML 模式文档中的 import 元素，用于从其他 WSDL 文档中导入 WSDL 定义；message 元素使用在 WSDL 文档的 type 元素中定义或在外部 WSDL 文档中定义的 XML 模式的内置类型、复杂类型或元素描述了消息的有效负载；portType 元素和 operation 元素描述了 Web 服务的接口并定义了它的方法。portType 元素和 operation 元素类似于 Java 接口和接口中定义的方法声明。operation 元素使用一个或者多个 message 类型来定义它的输入和输出的有效负载；Binding 元素将 portType 元素和 operation 元素赋给一个特殊的协议和编码样式；service 元素负责将 Internet 地址赋给一个具体的绑定。

3. UDDI

UDDI(Universal Description、Discovery and Integration，统一描述、发现和集成)是一套基于 Web 的、分布式的、为 Web 服务提供的信息注册中心的实现标准规范，同时也包含一组使企业能将自身提供的 Web 服务注册以使得别的企业能够发现的访问协议的实现标准。

UDDI 同时也是 Web 服务集成的一个体系框架。它包含了服务描述与发现的标准规范。UDDI 规范利用了 W3C 和 Internet 工程任务组织的很多标准作为其实现基础，例如 XML 语言、HTTP 和域名服务等协议。另外，在跨平台的设计特性中，UDDI 主要采用了 SOAP 进行通信。

UDDI 包含于完整的 Web 服务协议栈内，而且是协议栈基础的主要部件之一，支持创建、说明、发现和调用 Web 服务，如图 7-2 所示。

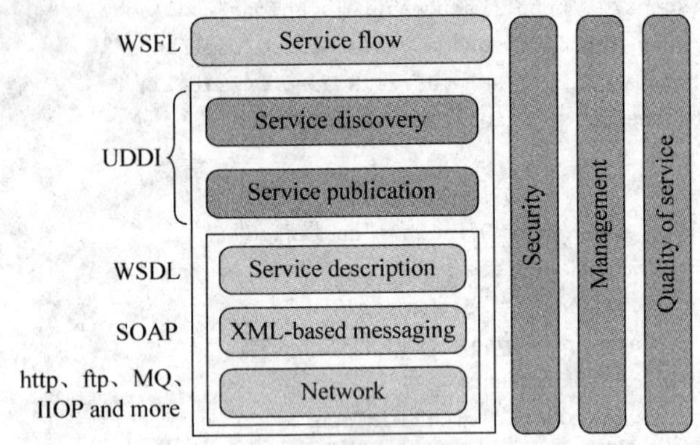

图 7-2　UDDI 的分层 Web 服务协议栈

UDDI 实现了一组可公开访问的接口，通过这些接口，网络服务可以向服务信息库注册其服务信息、服务需求者可以找到分散在世界各地的网络服务。

UDDI 的核心是其注册机制。UDDI 注册中心包含了通过程序手段可以访问到的对企业和企业支持的服务所做的描述。此外，还包含对 Web 服务所支持的因行业而异的规范、分类法定义(用于对于企业和服务很重要的类别)以及标识系统(用于对于企业很重要的标识)的引用。UDDI 提供了一种编程模型和模式，它定义与注册中心通信的规则。UDDI 规范中所有 API 都用 XML 来定义，包装在 SOAP 信封中，在 HTTP 上传输。UDDI 的工作原理如图 7-3 所示。

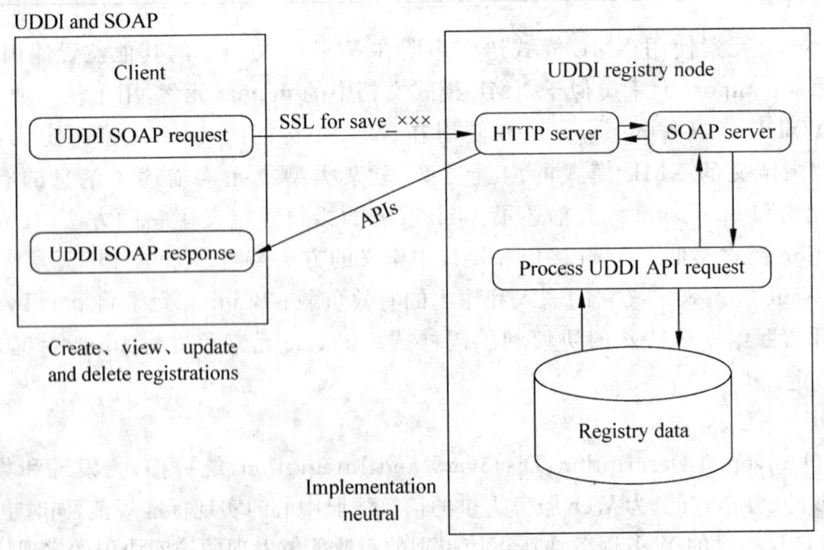

图 7-3　UDDI 消息在客户机和注册中心之间的流动

图 7-3 说明了 UDDI 消息的传输过程：通过 HTTP 从客户机的 SOAP 请求传到注册中心节点，然后再反向传输。注册中心服务器的 SOAP 服务器接收 UDDI SOAP 消息并进行处理，然后把 SOAP 响应返回给客户机。对注册中心来说，客户机发出的要修改数据的请求必须确保是安全的、经过验证的事务。

7.1.3　Ksoap2 编程

在 Android SDK 中并没有提供调用 Web 服务的库，因此需要使用第三方类库来调用 Web 服务。Google 为 Android 平台开发 Web 服务客户端提供了 ksoap2-android 项目[①]，它是 Android 平台上一个高效、轻量级的 SOAP 开发包，但这个项目并未直接集成在 Android 平台中，需要开发人员自行下载。

下面以一个查询电话号码归属地的 Web 服务为例，来分析 Android 中基于 Ksoap 编程的基本步骤。该 Web 服务终端如图 7-4 所示。

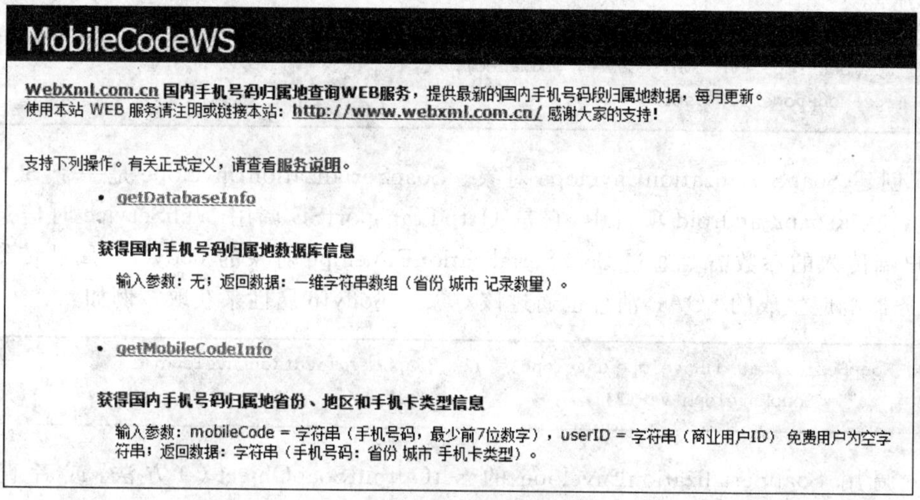

MobileCodeWS

WebXml.com.cn 国内手机号码归属地查询WEB服务，提供最新的国内手机号码段归属地数据，每月更新。
使用本站 WEB 服务请注明或链接本站：http://www.webxml.com.cn/ 感谢大家的支持！

支持下列操作。有关正式定义，请查看服务说明。

- getDatabaseInfo

  获得国内手机号码归属地数据库信息

  输入参数：无；返回数据：一维字符串数组（省份 城市 记录数量）。

- getMobileCodeInfo

  获得国内手机号码归属地省份、地区和手机卡类型信息

  输入参数：mobileCode = 字符串（手机号码，最少前7位数字），userID = 字符串（商业用户ID）免费用户为空字符串；返回数据：字符串（手机号码：省份 城市 手机卡类型）。

图 7-4　号码归属地 Web 服务端

单击其中的 getMobileCodeInfo 方法链接，在新打开页面的测试区域输入 mobileCode 参数，并单击“调用”按钮，显示如图 7-5 所示的响应 XML 数据信息。

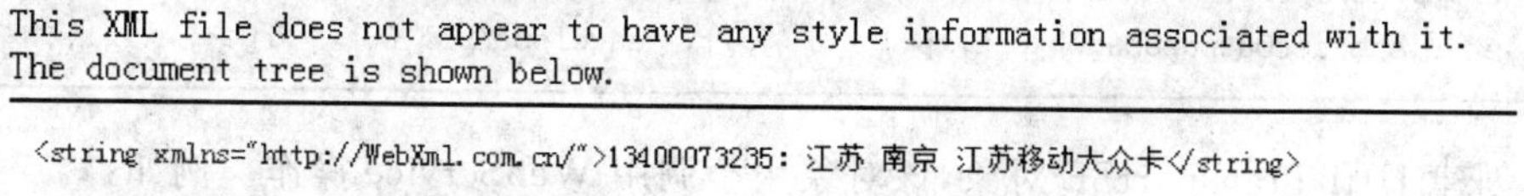

```
This XML file does not appear to have any style information associated with it.
The document tree is shown below.

<string xmlns="http://WebXml.com.cn/">13400073235：江苏 南京 江苏移动大众卡</string>
```

图 7-5　getMobileCodeInfo 方法响应信息

① 创建 Android 项目，并在项目的 libs 中导入 ksoap2-android 包（当前版本 ksoap2-android-assembly-3.4.0-jar-with-dependencies.jar）。

② 创建 Web 服务工具类，如 WebServiceUtil，并指定 Web 服务的命名空间、服务端点、调用方法名称。例如：

---

① 项目主页：https://code.google.com/p/ksoap2-android。

```
String nameSpace = "http://WebXml.com.cn/";
String methodName = "getMobileCodeInfo";
String endPoint =
        "http://webservice.webxml.com.cn/WebServices/MobileCodeWS.asmx";
```

③ 创建 SoapObject 对象,创建该对象时需要传入所要调用 WebService 的命名空间、WebService 方法名。例如:

```
SoapObject rpc = new SoapObject(nameSpace, methodName);
```

④ 如果有参数需要传给 WebService 服务器端,调用 SoapObject 对象的 addProperty (Stringname,Object value)方法来设置参数,该方法的 name 参数指定参数名;value 参数指定参数值。本例中,两个参数 mobileCode、userId 不可以随便写,必须和提供的参数名相同。例如:

```
rpc.addProperty("mobileCode", phoneSec);
rpc.addProperty("userId", "");
```

⑤ 创建 SoapSerializationEnvelope 对象。SoapSerializationEnvelope 是一个 SOAP 消息封包。在 ksoap2-android 项目中,它是 HttpTransportSE 调用 WebService 时信息的载体,客户端传入的参数需要通过 SoapSerializationEnvelope 对象的 bodyOut 属性传给服务器;服务器响应生成的 SOAP 消息也通过该对象的 bodyIn 属性来获取。例如:

```
SoapSerializationEnvelope envelope = new SoapSerializationEnvelope(
        SoapEnvelope.VER12);
```

⑥ 调用 SoapSerializationEnvelope 的 setOutputSoapObject()方法,或者直接对 bodyOut 属性赋值,将 SoapObject 对象设为 SoapSerializationEnvelope 的传出 SOAP 消息体。例如:

```
envelope.bodyOut = rpc;
envelope.dotNet = true;                //设置是否调用的是 dotNet 开发的 WebService
envelope.setOutputSoapObject(rpc);  //等价于 envelope.bodyOut = rpc
```

⑦ 创建 HttpTransportSE 对象,该对象用于调用 WebService 操作。例如:

```
HttpTransportSE transport = new HttpTransportSE(endPoint);
```

⑧ 调用 HttpTransportSE 对象的 call()方法,并以 SoapSerializationEnvelope 作为参数调用远程 Web 服务。例如:

```
try {
    transport.call(nameSpace + methodName, envelope);
} catch (Exception e) {
```

```
    e.printStackTrace();
}
```

⑨ 调用完成后,访问 SoapSerializationEnvelope 对象的 bodyln 属性,该属性返回一个 SoapObject 对象,该对象就代表了 Web 服务的返回消息。解析该 SoapObject 对象,即可获取调用 Web 服务的返回值。例如:

```
SoapObject object = (SoapObject) envelope.bodyIn;
String result = object.getProperty("getMobileCodeInfoResult").toString();
```

下面给出一个完整的获取天气预报信息的 Web 服务客户端工具类,代码如下:

```
public class WeatherWebServiceReader {
    public static final String SERVICE_NS = "http://WebXml.com.cn/";
    public static final String SERVICE_URL =
            "http://webservice.webxml.com.cn/WebServices/WeatherWS.asmx";

    /**
     * 此方法用于通过 WebService 获取省份列表,服务的方法名字为"getRegionProvince"
     *
     * @return List<String> 返回省份列表集合
     */
    public static ArrayList<String> getProvince() {
        ArrayList<String> provinces = null;
        String methodName = "getRegionProvince";
        HttpTransportSE ht = new HttpTransportSE(SERVICE_URL, 30000);
        SoapSerializationEnvelope envelope = new SoapSerializationEnvelope(
                SoapEnvelope.VER12);
        SoapObject soapRequest = new SoapObject(SERVICE_NS, methodName);
        envelope.bodyOut = soapRequest;
        envelope.dotNet = false;
        try {
            ht.call(SERVICE_NS + methodName, envelope);
            if (envelope.getResponse() != null) {
                SoapObject detail = (SoapObject) envelope.bodyIn;
                provinces = getProvinceOrCity(detail);
            }
        } catch (IOException e) {
            e.printStackTrace();
        } catch (XmlPullParserException e) {
            e.printStackTrace();
        }
        return provinces;

    }

    /**
     * 此方法用于根据 WebService 响应的服务解析出省份对象集合或者指定省份下属城市集合
```

```
     *
     * @param detail
     *            SoapObject 表示 WebService 服务响应的 SoapObject 对象
     * @return ArrayList<String> 返回值表示省份名称集合,或者城市名称集合
     */
    public static ArrayList<String> getProvinceOrCity(SoapObject detail) {
        ArrayList<String> result = new ArrayList<String>();
        System.out.println(detail.toString());
        String resultResponse = detail.toString();
        resultResponse = resultResponse.substring(
                resultResponse.indexOf("anyType{") + 8,
                resultResponse.lastIndexOf("};"));
        String[] nameList = resultResponse.split(" ");
        for (int i = 0; i < nameList.length; i++) {
            result.add(nameList[i].substring(nameList[i].indexOf("=") + 1,
                    nameList[i].indexOf(",")));
        }
        return result;
    }

    /**
     * 此方法用于通过向 WebService 发送省份名称获得该省市相关的城市集合
     * 服务方法名称为 getSupportCityString
     *
     * @param province
     *            String 表示指定省份名称
     * @return ArrayList<String> 指定省市相关的集合
     */
    public static ArrayList<String> getCitiesByProvinceName(
            String provinceName) {
        ArrayList<String> cities = null;
        String methodName = "getSupportCityString";
        HttpTransportSE ht = new HttpTransportSE(SERVICE_URL, 30000);
        SoapSerializationEnvelope envelope = new SoapSerializationEnvelope(
                SoapEnvelope.VER12);
        SoapObject soapRequest = new SoapObject(SERVICE_NS, methodName);
        soapRequest.addProperty("theRegionCode", provinceName);
        envelope.bodyOut = soapRequest;
        envelope.dotNet = false;
        try {
            ht.call(SERVICE_NS + methodName, envelope);
            if (envelope.getResponse() != null) {
                SoapObject detail = (SoapObject) envelope.bodyIn;
                cities = getProvinceOrCity(detail);
            }
        } catch (IOException e) {
            e.printStackTrace();
        } catch (XmlPullParserException e) {
            e.printStackTrace();
```

```
        }
        return cities;
    }

    /**
     * 此方法用于根据城市名称获得天气预报 WebService 提供的详细天气气象信息
     * 服务方法名称为 theCityCode
     *
     * @param cityName
     *            String 表示指定的城市名称
     * @return SoapObject 返回 SoapObject 对象表示城市气象信息的 Soap 消息对象
     */
    public static SoapObject getWeatherByCity(String cityName) {
        SoapObject weatherDetail = null;
        String methodName = "getWeather";
        HttpTransportSE ht = new HttpTransportSE(SERVICE_URL, 30000);
        SoapSerializationEnvelope envelope = new SoapSerializationEnvelope(
                SoapEnvelope.VER12);
        SoapObject soapRequest = new SoapObject(SERVICE_NS, methodName);
        soapRequest.addProperty("theCityCode", cityName);
        envelope.bodyOut = soapRequest;
        envelope.dotNet = false;
        try {
            ht.call(SERVICE_NS + methodName, envelope);
            if (envelope.getResponse() != null) {
                weatherDetail = (SoapObject) envelope.bodyIn;
            }
        } catch (IOException e) {
            e.printStackTrace();
        } catch (XmlPullParserException e) {
            e.printStackTrace();
        }
        return weatherDetail;
    }
}
```

7.2　开放接口编程

开放平台是互联网发展的新趋势，其建设思想在于将网站提供的服务封装成一系列对外开放的计算机编程数据接口，提供给第三方开发者使用，第三方开发者通过这些开放接口开发丰富多彩的应用，并以此获得可观的社会效益和经济利益。

7.2.1　开放平台概述

在互联网时代，把网站的服务封装成一系列计算机易识别的数据接口开放出去，供第三方开发者使用，这种行为就叫作开放 API，提供开放 API 的平台被称为开放平台。通过开放平台,网站不仅能提供对 Web 网页的简单访问,还可以进行复杂的数据交互,将它们的 Web 网站转换为与操作系统等价的开发平台。

Facebook 的成功展示了开放平台的巨大潜力。从 2008 年开始,人人网和淘宝网推出自己的开放平台计划。到 2010 年,新浪微博、百度、盛大、开心网、腾讯、360 等国内互联网企业相继尝试开放部分自己产品服务与数据的 API,展开与第三方开发者的争夺战。

国内主流开放平台积极布局 SNS(Social Network Site,即社交网站)、搜索、电商、桌面、微博。人人网发展社区应用、新浪微博主打内容开放、淘宝网发展电子商务应用、腾讯发展多维度平台、百度主打 Web 应用、360 发展多客户端应用等。

开放平台实现了应用开放、横向开放和数据开放。应用开放是开放自身平台的各种标准接口,欢迎第三方提供各类应用,共享用户,共同服务。横向开放是开放平台本身欢迎第三方平台或网站互联互通,让用户在不同平台和网站间畅通无阻。数据开放是在保护用户隐私前提下,开放用户基本数据、关系数据和行为数据,同第三方一起打造个性化、个人化、智能化、实时化的服务模式。

开放平台运营商根据核心业务定位确定开放尺度。在开放平台大潮中,互联网企业开放到何种程度,开放政策如何制定,开放底线在哪里,归根结底取决于企业的业务定位以及核心竞争力。

盈利模式决定了平台是否开放以及开放的程度。一类开放平台自己不创造内容,它们是真正的平台角色,让信息供给者、广告供给者、信息消费者在自己的服务中云集,自己从中获取收益。另一类开放平台利用互联网增值服务,包括网游和虚拟物品售卖等盈利,开放是为了利用开发者贡献的内容吸引更多用户,增加用户黏性,从而进一步巩固在核心业务的控制地位,保持长期盈利的能力。

作为开发者,首要考虑的还是接口用户需求和收益分成,对于不同的开放平台,开发者和服务商之间的分成没有统一的分配。

7.2.2 OAuth 授权

为了安全地访问在线服务,用户需要在服务上进行身份验证,即要提供其身份证明。对于一个要访问第三方服务的程序来说,安全问题甚至更复杂。不仅仅是用户需要在访问服务前进行身份验证,而且程序也要进行身份验证来授权用户。

1. OAuth 与 OAuth 2.0

OAuth 是一个分布式身份验证和授权的开放标准,它于 2006 年由 Twitter 和业务合作伙伴 Ma.gnolia 开发,用来方便地创建一些桌面小部件,这些小部件允许第三方应用在用户授权的情况下访问其在网站上存储的信息资源,而这一过程中网站无须将用户的账号密码告诉第三方应用。

OAuth 的设计思路是在“客户端”与“服务提供商”之间设置一个授权层(authorization layer)。“客户端”不能直接登录“服务提供商”,只能登录授权层,以此将用户与客户端区分开来。“客户端”登录授权层所用的令牌(token),与用户的密码不同。用户可以在登录时,指定授权层令牌的权限范围和有效期。“客户端”登录授权层以后,“服务提供商”根据令牌的权限范围和有效期,向“客户端”开放用户储存的资料。

OAuth 2.0 是 OAuth 协议的当前版本。OAuth 2.0 提供一个单值,叫作认证令牌

(auth token)，代表用户身份和程序身份验证授权。OAuth 2.0 针对 1.0 版本存在的各种问题提出了如下解决方案：

- OAuth 2.0 提出了多种流程，各个客户端按照实际情况选择不同的流程来获取 access_token。这样就解决了对移动设备等第三方的支持，也解决了拓展性的问题。
- OAuth 2.0 删除了烦琐的加密算法。利用 https 传输保证了认证的安全性。
- OAuth 2.0 的认证流程一般只有两步，对开发者来说减轻了其负担。
- OAuth 2.0 提出了 access_token 的更新方案，获取 access_token 的同时也获取 refresh_token。access_token 是有过期时间的，refresh_token 的过期时间较长，这样能随时使用 refresh_token 对 access_token 进行更新。

图 7-6 为人人网基于 OAuth 2.0 实现授权的示例。

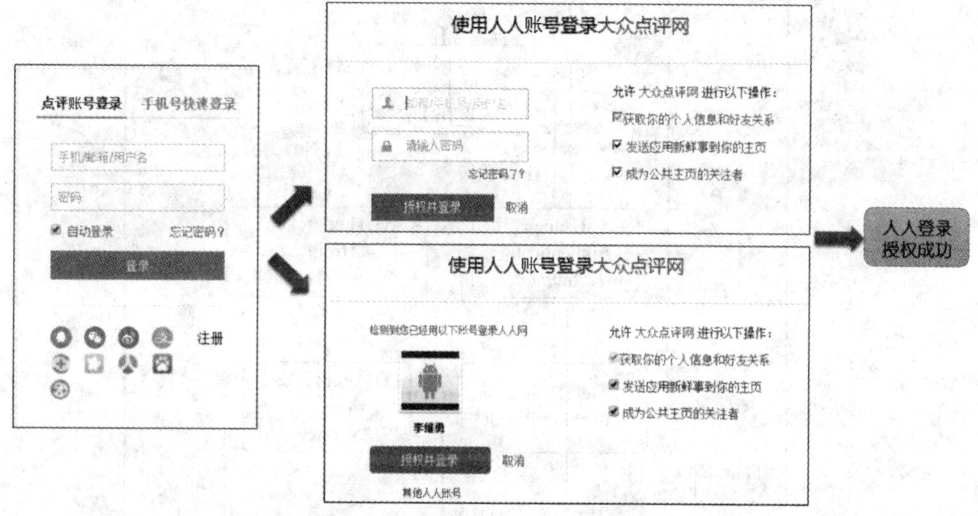

图 7-6　人人网 OAuth 2.0 授权示例

2. Android 中的 OAuth 授权流程

OAuth 2.0 的授权流程一般为：

① 用户打开客户端以后，客户端要求用户给予授权。

② 用户同意给客户端授权。

③ 客户端使用上一步获得的授权，向认证服务器申请令牌。

④ 认证服务器对客户端进行认证以后，确认无误，同意发放令牌。

⑤ 客户端使用令牌，向资源服务器申请获取资源。

⑥ 资源服务器确认令牌无误，同意向客户端开放资源。

图 7-7 演示了 Android 平台向 Google 服务器请求一个验证令牌的流程：

为了得到一个验证令牌，首先需要在 AndroidManifest 文件中添加 ACCOUNT_MANAGER 权限(为了访问网络，也必须添加 INTERNET 权限)，例如：

```
<uses-permission android:name="android.permission.ACCOUNT_MANAGER" />
<uses-permission android:name="android.permission.INTERNET" />
```

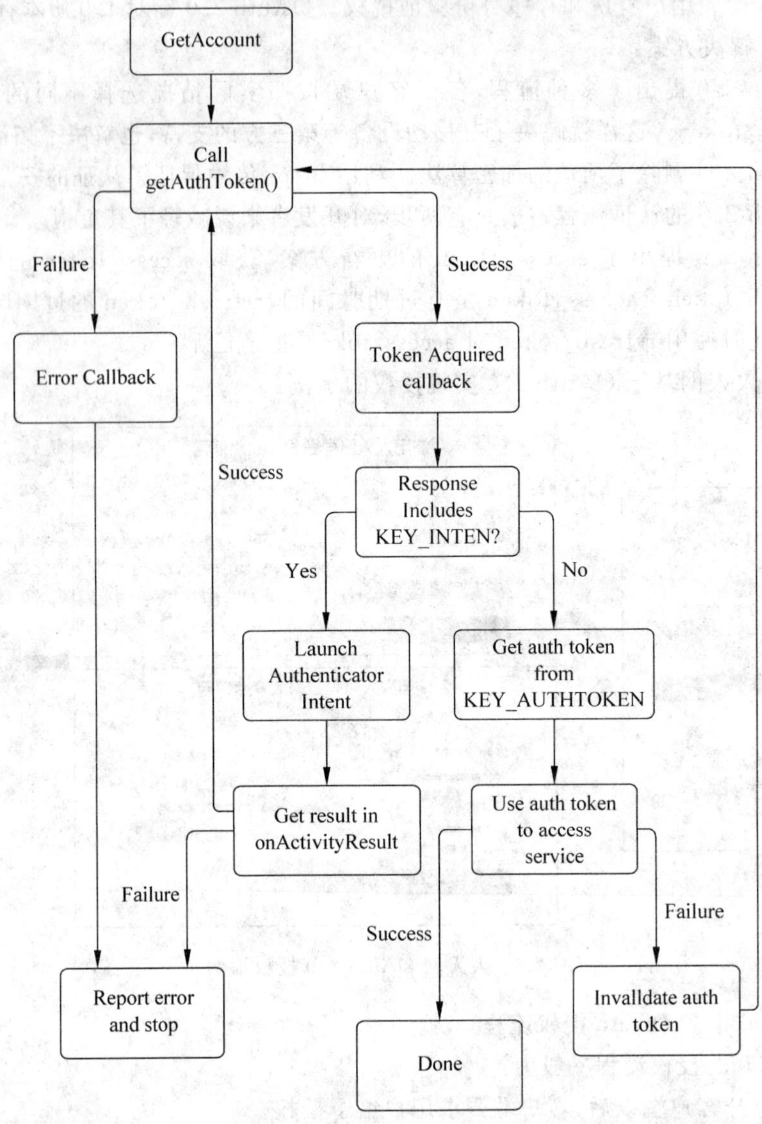

图 7-7　获取令牌流程

调用 AccountManager.getAuthToken()方法来获取令牌。由于获取令牌的过程需要设计网络访问,因此,AccountManager 中的方法都是异步的,并且获得令牌的过程往往通过多个回调方法来完成,例如:

```
AccountManager am = AccountManager.get(this);
Bundle options = new Bundle();

am.getAuthToken(
    myAccount_,                          //用 getAccountsByType()来检索的账户
    "Manage your tasks",                 //令牌范围
    options,                             //特殊验证选项
    this,                                //Activity
```

```
    new OnTokenAcquired(),                    //成功获取令牌后调用的回调方法
    new Handler(new OnError()));              //错误发生时调用的回调方法
```

其中，OnTokenAcquired 是一个继承自 AccountManagerCallback 的类。AccountManager 在 OnTokenAcquired 中调用 run( )方法，该方法需要传递一个含有一个 Bundle 的 AccountManagerFuture 实例。如果调用成功，那么令牌中就会包含这个 Bundle。

下面的代码演示了从 Bundle 中获取令牌的方法：

```
private class OnTokenAcquired implements AccountManagerCallback<Bundle> {
    @Override
    public void run(AccountManagerFuture<Bundle> result) {
        //Get the result of the operation from the AccountManagerFuture.
        Bundle bundle = result.getResult();

        //The token is a named value in the bundle. The name of the value
        //is stored in the constant AccountManager.KEY_AUTHTOKEN.
        token = bundle.getString(AccountManager.KEY_AUTHTOKEN);
        ...
    }
}
```

7.2.3 人人网编程

人人网开放平台(http://dev.renren.com/)提供了一种新的接口调用方式，允许被人人网用户授予权限的第三方应用以人人网用户的身份来读写人人网的资源(例如用户基本资料、好友关系、照片等)。

使用人人网账号登录移动客户端，并利用人人网开放平台提供的社交图谱(Social Graph)和传播渠道，增进用户与好友的交互，提升使用体验，并获得广泛传播。

Android 手机客户端接入人人网，可以有两种实现方式：一种是直接使用人人网开放平台提供的各种接口，如用作验证和授权的 OAuth 2.0，提供数据的底层 Rest API，以及嵌入各种 Widget；另一种是使用人人网开放平台官方封装的开源 Android SDK。

1. 创建应用

创建人人网应用首先需要一个人人网账号，如果没有人人网账户，可以先去注册一个。登录后输入 dev.renren.com 进入人人网开放平台页面，点击"我的应用"进入应用管理中心。这时需要进行开发者身份认证，填写完开发者认证信息后，开放平台会给注册邮箱发送一封认证邮件。点击认证链接后完成开发者身份认证返回应用中心。这时就是一名经过认证的开发者了。

成为一个开发者后，可以按照图 7-8 的流程使用人人网开发平台。

点击界面上的"创建应用"按钮，这时会进入基本信息页面，需要正确并完整地填写基本信息之后，才可以创建应用，如图 7-9 所示。

按照界面提示步骤，一步一步完成应用的创建后，将获得 APP ID、API KEY 和 Secret Key。

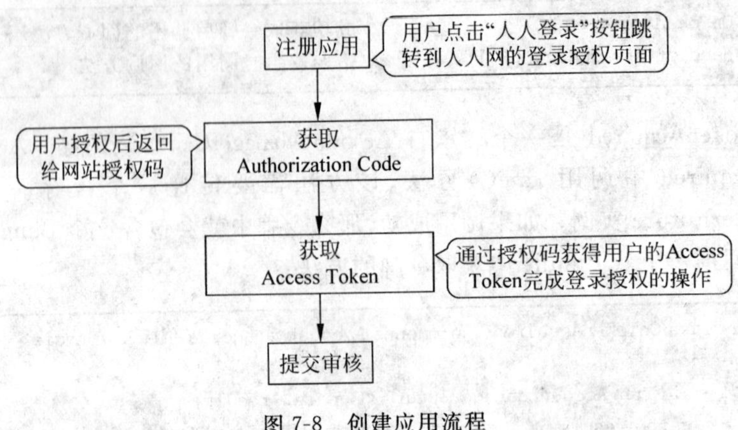

图 7-8 创建应用流程

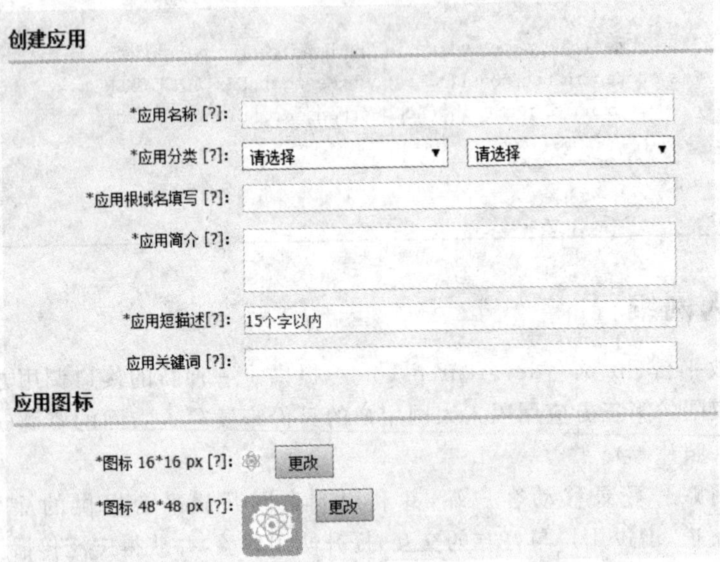

创建应用

*应用名称 [?]:

*应用分类 [?]: 请选择 请选择

*应用根域名填写 [?]:

*应用简介 [?]:

*应用短描述[?]: 15个字以内

应用关键词 [?]:

应用图标

*图标 16*16 px [?]: 更改

*图标 48*48 px [?]: 更改

图 7-9 创建应用界面

2. 基于 SDK 开发应用

本节使用人人网 Android SDK 来实现发布一张手机图片到人人网的功能。有关 Android 手机客户端应用接入方式的详细信息可以参考人人网的维基(http://wiki.dev.renren.com/wiki/移动客户端接入)。

下面介绍通过自定义的图片上传服务 UploadPictureService 文件实现图片上传的功能。步骤如下:

① 在 Eclipse 中导入从 https://bitbucket.org/renren_platform/下载的 SDK 项目 renren_android_connect。

② 在应用项目的 Properties 对话框中添加 renren_android_connect Library。

注意:在 Windows 系统下,library project 必须和 project 处于相同的盘符中,因为如果用了不同的盘符,project.properties 中的 android.library.reference.1 值变成绝对路径,而 ADT 推荐的是在 ubuntu 下使用,对 Windows 绝对路径的支持有错误。

③ 在上传图片的 Activity 的 onCreate()方法中初始化 Renren 对象，开发者可以使用该类调用人人网提供的网络接口。例如：

```
renren = new Renren(API_KEY, SECRET_KEY, APP_ID, this);
```

其中的 API_KEY、SECRET_KEY 和 APP_ID 即为创建应用时获取的信息。

④ 创建 AsyncTask 实现类异步处理图片上传，核心代码如下：

```
class UploadTask extends AsyncTask<String, Integer, String> {

    @Override
    protected String doInBackground(String... params) {
        try {
            Looper.prepare();
            renren.authorize(RenRenActivity.this, renRenAuthListener);
            Looper.loop();
        } catch (Exception e) {
            e.printStackTrace();
        }
        return null;
    }

    ...
}
```

⑤ 实现监听认证和授权动作的 RenrenAuthListener，代码如下：

```
private RenrenAuthListener renRenAuthListener = new RenrenAuthListener() {
    @Override
    public void onComplete(Bundle values) {
        Log.d(TAG, "Auth ok.");
        //待上传图片的路径
        final String imagePath = getImageAbsolutePath(uri);
        try {
            AlbumGetResponseBean beans = renren.getAlbums(null);
            if (beans == null) {
                noRenRenAlbums(imagePath);
            } else {
                List<AlbumBean> beanList = beans.getAlbums();
                showRenRenAlbums(beanList, imagePath);
            }
        } catch (RenrenException e) {
            e.printStackTrace();
        } catch (Throwable e) {
            e.printStackTrace();
        }

    }
```

```
22
23      @Override
24      public void onRenrenAuthError(RenrenAuthError renrenAuthError) {
25          renRenHandler.post(new Runnable() {
26              @Override
27              public void run() {
28                  Toast.makeText(RenRenActivity.this, "认证失败",
29                          Toast.LENGTH_SHORT).show();
30              }
31          });
32      }
33
34      @Override
35      public void onCancelLogin() {
36      }
37
38      @Override
39      public void onCancelAuth(Bundle values) {
40      }
41  };
```

其中,08 行的 AlbumGetResponseBean 封装了 List＜AlbumBean＞,用于返回相册列表(图 7-10)。如果没有相册列表,则将当前图片上传到默认目录,否则显示相册列表。

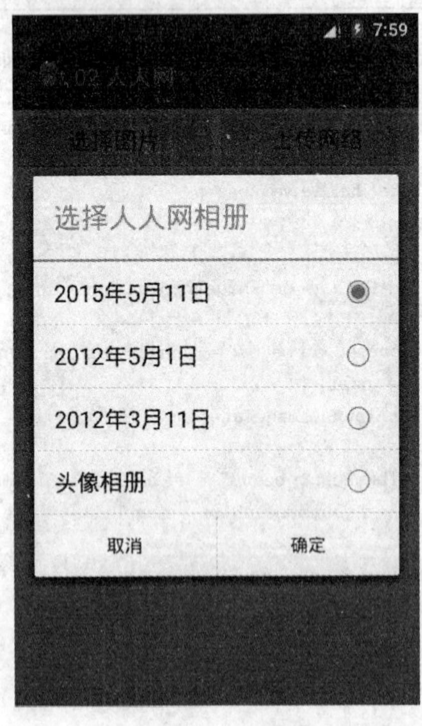

图 7-10　相册列表

⑥ 实现默认相册列表的上传，代码如下：

```
private void noRenRenAlbums(final String imagePath) {
    AlertDialog.Builder builder = new AlertDialog.Builder(
            RenRenActivity.this);
    builder.setTitle("注意");
    builder.setMessage("上传到默认相册?");
    builder.setPositiveButton("确定",
            new DialogInterface.OnClickListener() {
        public void onClick(DialogInterface dialog, int id) {
            try {
                publishPhoto(-1, imagePath);
            } catch (Throwable e) {
                Toast.makeText(getApplicationContext(), "图片上传失败",
                        Toast.LENGTH_SHORT).show();
                e.printStackTrace();
            }
        }
    });
    builder.setNegativeButton("取消",
            new DialogInterface.OnClickListener() {
        public void onClick(DialogInterface dialog, int id) {
        }
    });
    AlertDialog alert = builder.create();
    alert.show();
}
```

⑦ 实现指定相册的上传，代码如下：

```
private void showRenRenAlbums(final List<AlbumBean> beans,
        final String imagePath) {
    int size = beans.size();
    final String albums[] = new String[size];
    for (int i = 0; i < size; i++) {
        albums[i] = beans.get(i).getName();
    }
    selectedItem = 0;
    AlertDialog.Builder builder = new AlertDialog.Builder(this);
    builder.setTitle("选择人人网相册");
    builder.setSingleChoiceItems(albums, 0,
            new DialogInterface.OnClickListener() {
                public void onClick(DialogInterface dialog, int item) {
                    selectedItem = item;
                }
            });
    builder.setPositiveButton("确定",
            new DialogInterface.OnClickListener() {
        public void onClick(DialogInterface dialog, int id) {
```

```
            try {
                publishPhoto(beans.get(selectedItem).getAid(), imagePath);
            } catch (Throwable e) {
                Toast.makeText(getApplicationContext(), "图片上传失败",
                        Toast.LENGTH_SHORT).show();
                e.printStackTrace();
            }
        }
    });
    builder.setNegativeButton("取消",
            new DialogInterface.OnClickListener() {
        public void onClick(DialogInterface dialog, int id) {
        }
    });
    AlertDialog alert = builder.create();
    alert.show();
}
```

⑧ 实现图片上传的方法 publishPhoto(),代码如下：

```
private void publishPhoto(long aid, String picPath) throws RenrenException,
        Throwable {

    PhotoUploadRequestParam param;

    File f = new File(picPath);

    if (f.length() > 2097152) {
        Bitmap resizeBmp = null;
        BitmapFactory.Options opts = new BitmapFactory.Options();
        opts.inSampleSize = 10;
        resizeBmp = BitmapFactory.decodeFile(f.getPath(), opts);
        String extension = picPath.substring(picPath.lastIndexOf(".") + 1,
                picPath.length()).toLowerCase();
        saveBitmapToSDCard(resizeBmp, "/sdcard/_tmp." + extension);
        param = new PhotoUploadRequestParam(new File("/sdcard/_tmp."
                + extension));
    } else {
        param = new PhotoUploadRequestParam(new File(picPath));
    }

    if (aid != -1) {
        param.setAid(aid);
    }
    if (renren.publishPhoto(param) == null) {
        Toast.makeText(getApplicationContext(), "图片上传失败",
                Toast.LENGTH_SHORT).show();
    } else {
        Toast.makeText(getApplicationContext(), "图片上传成功",
```

```
30                  Toast.LENGTH_SHORT).show();
31      }
32  }
```

其中，04 行的 PhotoUploadRequestParam 是上传图片 API 的请求参数 bean。08～18 行是对大尺寸照片的处理技巧，避免 OOM 异常。

7.2.4　新浪微博编程

新浪微博开放平台(http://open.weibo.com/index.php)为方便移动应用接入微博，微博平台提供了相关接口及个性化的产品结合模式，并不断优化微博移动端解决方案，提供更多定制化、个性化服务，满足了多元化移动终端用户随时随地快速登录、分享信息的需求，助力实现移动 Apps、健康设备、智能家居、车载等多类型终端的社会化接入。

1. 接入平台

接入平台的整体流程如图 7-11 所示。

图 7-11　SDK 接入流程

(1) 注册成为开发者，创建移动应用。

如果还不是一名开发者，先注册成为开发者(具体参考：http://open.weibo.com/wiki/%E6%96%B0%E6%89%8B%E6%8C%87%E5%8D%97)。

在开发者页面单击“登录”或者“创建应用”，通过账号登录成为一名开发者。一个微博账号可以管理 10 个不同的应用，建议开发人员使用官方微博的账号，以便统一管理。

单击“创建应用”，即进入目标应用的类型选择环节。创建应用时，开发者需要谨慎选择应用对应平台，不同的平台建议使用不同的 APP KEY 开发。

然后在开发者信息设置页填写真实资料，成为微博认证的开发者，这个过程需要通过邮箱验证和手机验证。

(2) 获取授权。

创建应用完成后，可以在“我的应用-应用信息”页面中查看所创建应用的 APP KEY 及 APP SECRET，这些都是调用微博开放平台各 API 的身份标志。

在“我的应用-应用信息-高级信息”页面中填写应用回调页(OAuth2.0 客户端默认回调页是 https://api.weibo.com/oauth2/default.html)，这样才能使 OAuth2.0 授权正常进行。如果 APP SECRET 发生泄露，也可以通过该页面中的“重置”按钮对其重置。

在“我的应用-应用信息”页面中填写应用的平台信息，如图 7-12 所示。

Android 应用填写包名、签名及下载地址(关于各字段含义在控制台中均有说明)。

2. 基于 SDK 开发应用

本节以 sinaBlog① 的实现为例，介绍基于新浪微博开放平台的移动应用开发技术。

① 项目主页：https://github.com/NickAndroid/sinaBlog。

应用基本信息

应用类型: 普通应用 - 客户端

应用名称: 测试安卓应用　该名称也用于来源显示，不超过10个汉字或20个字母

应用平台: 手机

iPhone　Android　BlackBerry

Windows Phone　Symbian

WebOS　Other

• Android包名:

• Android签名:　通过下载使用平台提供的签名工具获取

• Android下载地址:

图 7-12　应用基本信息

在介绍具体开发步骤前,先了解几个微博核心 API 的对象。

- Weibo: 微博 API 接口类,对外提供 Weibo API 的调用,包括登录、API 调用、微博分享等功能。
- AsyncWeiboRunner: 微博 API 异步执行类,封装了回调接口,通过创建线程来调用 Weibo 中的接口方法。
- Utility: 互联网工具类,包括接口请求 GET/POST 封装、BASE64 等 encode、decode 方法。
- WeiboException: 微博异常封装类,封装了微博的各个异常。

① 在 Eclipse 中导入微博官方 SDK(下载地址为 https://github.com/sinaweibosdk/weibo_android_sdk)。

② 在应用项目的 Properties 对话框中添加 WeiboSDK。

③ 在闪屏 Activity 的 onCreate()方法中初始化 WeiboAuth 对象。例如:

```
mWeiboAuth = new WeiboAuth(this, com.bpok.sina.sdk.Constants.APP_KEY,
        Constants.REDIRECT_URL, Constants.SCOPE);
```

其中的 APP_KEY 即为创建应用时获取的信息,REDIRECT_URL 是 OAuth2.0 客户端默认回调页,SCOPE 是 OAuth2.0 授权机制中 authorize 接口的一个参数(具体参见 http://open.weibo.com/wiki/Scope#scope.E8.AF.B4.E6.98.8E)。

有时为了自动保存登录信息,需要在 SharedPreferences 中存储上次已保存好的 AccessToken 等信息。例如:

```
Oauth2AccessToken mAccessToken;
mAccessToken = AccessTokenKeeper.readAccessToken(this);
```

其中,OAuth2AccessToken 是封装 Tokens 的属性类,继承自 Token,包含 Access_

token、OAuth_token_secret 等多个属性。

④ 创建登录 Activity，核心代码如下：

```
public class Login extends Activity {
    ...

    @Override
    protected void onCreate(Bundle savedInstanceState) {
        super.onCreate(savedInstanceState);
        this.requestWindowFeature(Window.FEATURE_NO_TITLE);
        setContentView(R.layout.activity_login);
        initLoginViews();
    }

    private void initLoginViews() {
        ...
        loginButton.setOnClickListener(new OnClickListener() {

            @Override
            public void onClick(View arg0) {
                mSsoHandler = new SsoHandler(Login.this, mWeiboAuth);
                mSsoHandler.authorize(new AuthListener());
            }
        });
    }

    class AuthListener implements WeiboAuthListener {

        @Override
        public void onComplete(Bundle values) {
            mAccessToken = Oauth2AccessToken.parseAccessToken(values);
            if (mAccessToken.isSessionValid()) {
                updateTokenView(false);
                AccessTokenKeeper.writeAccessToken(Login.this,
                        mAccessToken);
                ...
            } else {
                ...
            }
        }

    }

    @SuppressLint("SimpleDateFormat")
    private void updateTokenView(boolean hasExisted) {
        String date = new SimpleDateFormat("yyyy/MM/dd HH:mm:ss")
                .format(new java.util.Date(mAccessToken.getExpiresTime()));
        String format = getString(
                R.string.weibosdk_demo_token_to_string_format_1);
```

```
47          String message = String.format(format, mAccessToken.getToken(),
48                  date);
49          if (hasExisted) {
50              message = getString(R.string.weibosdk_demo_token_has_existed)
51                      + "\n" + message;
52          }
53      }
54
55  }
```

其中,24～39 行定义的 AuthListener 类是微博认证授权的回调类。SSO 授权时,需要在 onActivityResult()方法中调用 SsoHandler 的 authorizeCallBack()后,该回调才会被执行。当授权成功后,保存该 access_token、expires_in、uid 等信息到 SharedPreferences 中。

⑤ 实现微博主界面(如图 7-13 所示)。微博主界面主要是初始化用户信息、侧边菜单以及微博内容区。下面给出内容区域的 Fragment 实例 FragmentHome 的核心代码:

图 7-13 微博主界面

```
public class FragmentHome extends FragmentBase implements
        ActionBar.OnNavigationListener {
    ...

    @Override
    public View onCreateView(LayoutInflater inflater, ViewGroup container,
```

```
            Bundle savedInstanceState) {
        configManager = new ConfigManager(getActivity());
        if (configManager.getThemeMod() == 0)
            mRootView = inflater.inflate(R.layout.fragment_home, null);
        else {
            mRootView = inflater.inflate(R.layout.fragment_home_night, null);
        }
        readyHandler.sendEmptyMessageDelayed(0, 2000);
        initActionBar(inflater);
        this.requestRefershImpl = (RequestRefershImpl) getActivity();
        return mRootView;
    }

    private void initApi() {
        mAccessToken = AccessTokenKeeper.readAccessToken(mContext);
        mStatusesAPI = new StatusesAPI(mAccessToken);
    }

    private void getNewStatues(int featureType) {
        requestRefershImpl.onRequestRefersh();
        if (mStatusesAPI == null) {
            initApi();
        }
        mStatusesAPI.friendsTimeline(0L, 0L, page_status_count, 1, false,
                featureType, false, mListener);
        new GetAllStatusAsyncTask().execute();
        this.current_since_id = page_status_count;
    }

    private void updateListData(StatusList statuses) {
        if (statuses != null && statuses.total_number > 0) {
            if (mHomeListAdapter == null) {
                mHomeListAdapter = new HomeListAdapter(this, statuses,
                        getActivity());
                if (swingBottomInAnimationAdapter == null)
                    swingBottomInAnimationAdapter = new
                            SwingBottomInAnimationAdapter(
                                    mHomeListAdapter);
                swingBottomInAnimationAdapter.setListView(mListView);
                if (mListView == null) {
                    mListView = (UpDownRefershListView) mRootView
                            .findViewById(R.id.lv_home);
                }
                mListView.setAdapter(mHomeListAdapter);
            } else {
                mHomeListAdapter.statusList = statuses.statusList;
                mHomeListAdapter.notifyDataSetChanged();
            }
        }
```

```
        mListView.onPulldownRefreshComplete();
        requestRefershImpl.onRefershComplete();
        this.currentList = statuses;
    }

    private RequestListener mListener = new RequestListener() {
        @Override
        public void onComplete(String response) {
            if (!TextUtils.isEmpty(response)) {
                LogUtil.i(TAG, response);
                if (response.startsWith("{\"statuses\"")) {
                    StatusList statuses = StatusList.parse(response);
                    if (statuses != null && statuses.total_number > 0) {
                        cacheDataToStorge(response, "response.json");
                        updateListData(statuses);
                        currentList = statuses;
                    }
                } else if (response.startsWith("{\"created_at\"")) {
                    Status status = Status.parse(response);
                } else {
                    Log.i(TAG, response);
                }
            }
            requestRefershImpl.onRefershComplete();
        }

        @Override
        public void onWeiboException(WeiboException e) {
            LogUtil.e(TAG, e.getMessage());
            requestRefershImpl.onRefershComplete();
            ErrorInfo info = ErrorInfo.parse(e.getMessage());
            Log.i(TAG, info.toString());
        }
    };

    private RequestListener mListener_all_status = new RequestListener() {
        @Override
        public void onComplete(String response) {
            if (!TextUtils.isEmpty(response)) {
                LogUtil.i(TAG, response);
                if (response.startsWith("{\"statuses\"")) {
                    StatusList statuses = StatusList.parse(response);
                    if (statuses != null && statuses.total_number > 0) {
                        cacheDataToStorge(response, "all_response.json");
                        if (configManager.getFeatureType() == 0)
                            allList = statuses;
                    }
                } else {
                    Log.i(TAG, response);
```

```
105                 }
106             }
107             requestRefershImpl.onRefershComplete();
108         }
109 
110         @Override
111         public void onWeiboException(WeiboException e) {
112             LogUtil.e(TAG, e.getMessage());
113             requestRefershImpl.onRefershComplete();
114             ErrorInfo info = ErrorInfo.parse(e.getMessage());
115             Log.i(TAG, info.toString());
116         }
117     };
118 
119 }
```

其中，25～34 行定义的 getNewStatues()方法主要用于下拉刷新时获取最新动态。StatusesAPI(Oauth2AccessTokenaccesssToken)封装了微博的相关信息(详细 API 信息参见 http://open.weibo.com/wiki/%E5%BE%AE%E5%8D%9AAPI#.E5.BE.AE.E5.8D.9A)，提供的方法包括以下几种。

- update()：分享文字和图片到微博中。
- public_timeline()：获取最新的公共微博。
- friends_timeline()：获取当前登录用户及其所关注用户的最新微博。
- home_timeline()：获取当前登录用户及其所关注(授权)用户的最新微博。
- user_timeline()：获取用户发布的微博。

36～59 行定义的 updateListData()方法用于刷新后更新微博列表信息。61～89 行定义的 RequestListener 实例是微博 API 的回调接口，通过调用 StatusList.parse(response)方法可以解析出微博信息。

除了上面介绍的 Activity 之外，还可以实现微博撰写界面、回复界面、用户信息界面等。

7.3　习　　题

1. 编程实现基于 http://fy.webxml.com.cn/webservices/EnglishChinese.asmx 的在线翻译功能。

2. 编程实现基于百度云的文件上传功能。

第 8 章　Google 云服务

Android 框架通过提供丰富的网络连接的 API,来帮助构建丰富的基于云端的应用,并使这些应用可以把数据同步到一个远程服务器上,同时确保多个设备保持同步,并将宝贵数据备份到云端。

8.1　Google 云备份

Google 为了给 Android 2.2 版本以上的应用程序的数据和配置信息提供数据还原点,在 Android 的数据备份框架中提供了把 Android 的数据复制到远程云存储的 API 接口。这样,在用户恢复出厂设置或者更换新的 Android 设备时,Google 云将对再次安装的应用程序自动恢复备份数据,整个过程对于用户而言完全透明。Google 云备份完善了程序的功能,提高了用户的体验。Android 4.0 以后版本增加了所有应用(包括系统自带的应用)的完全备份和恢复功能。

8.1.1　注册 Android 备份服务

只有应用程序注册了 Android 的备份服务,才会被允许用这个服务来备份和恢复数据。如果应用程序想要备份数据,那么为了使用 Android 的备份服务,就必须注册备份服务密钥,并把它包含在应用的 Android 清单文件中。当 Android 的备份管理器开始为应用程序备份或恢复数据时,Android 备份服务的备份传输器会检查清单文件中的备份服务密钥,只有这个密钥有效,它才会继续执行备份或恢复数据的操作。

注册备份服务密钥的步骤如下:

① 登录 https://developer.android.com/google/backup/signup.html?csw=1,选中下方的 I have read and agree with the Android Backup Service Terms of Service 复选框,然后在 Application package name 文本框中输入应用的包的名称,再单击 Register with Android Backup Service 按钮。

② 成功注册本服务后,会收到一个备份服务密钥和对应的<meta-data> XML 代码。例如:

```
Android Backup Service Key
Thank you for registering for an Android Backup Service Key!
Your key is:
AEdPqrEAAAAI679CqOut5LOOvfH60cqdyWz7TqhvQkkHdXN4ng
This key is good for the app with the package name:
cn.liwy.project08
```

```
Provide this key in your AndroidManifest.xml file with the following <meta-data> element,
placed inside the <application> element:
<meta-data android:name="com.google.android.backup.api_key" android:value="
AEdPqrEAAAAI679CqOut5LOOvfH60cqdyWz7TqhvQkkHdXN4ng" />
For more information, see the Backup Dev Guide, or go back to the Android Backup Service
registration.
```

必须把这段 XML 代码放在 AndroidManifest.xml 的<application>元素下面作为它的子节点。例如：

```
<application android:label="MyApplication"
      android:backupAgent="MyBackupAgent">
   ...
   <meta-data android:name="com.google.android.backup.api_key"
      android:value="AEdPqrEAAAAI679CqOut5LOOvfH60cqdyWz7TqhvQkkHdXN4ng"
   />
</application>
```

其中，android:name 的值必须是 com.google.android.backup.api_key，且 android:value 必须是从 Android 备份服务注册收到的备份服务密钥。android:backupAgent 的值是应用程序的备份代理类。

当使用 Android 备份服务的设备运行这个应用程序时，系统会确认备份服务密钥的有效性。如果有效，Android 备份服务会使用设备上的主 Google 账号把用户的数据保存到 Google 服务器上。

注意：每一个备份密钥只在指定的包名下有效，如果有不同的应用，必须用各自的包名称作为它们注册独立的备份服务密钥。

由 Android 备份服务提供的备份传输系统不保证所有基于 Android 的设备都支持备份功能。一些设备可能使用不同的传输系统支持备份功能，另外一些可能根本不支持。对于应用程序，没有一种方法能够知道设备使用的是哪一种传输系统。然而，如果应用实现了备份功能，那么总是应当包含一个备份服务密钥的。当设备使用 Android 备份服务传输系统时，这个密钥用于 Android 备份服务，使应用程序能够执行备份。如果设备不适用 Android 备份服务，那么<meta-data>元素和备份服务密钥会被忽略。

小贴士：

对于保护用户备份数据的安全性，Google 有着清醒的责任认识。为了提供备份和恢复功能，Google 提供了安全的备份数据传输机制。Google 会按照它的隐私政策来对待个人信息。另外，用户能够通过 Android 系统的隐私设置来禁用数据备份功能。当备份功能被禁用时，Android 备份服务会删除所有的被保存的备份数据。用户能够重新启用设备的备份功能，但 Android 的备份服务器不会恢复之前的任何被删除的数据。

8.1.2　备份管理器

在数据备份的过程中，Android 备份管理器(BackupManager)的主要任务是查找应用程序中需备份的数据，并把数据交给备份传输器，传输器再把数据传送给云存储。在数据恢复时,备份管理器负责从备份传输器取回备份数据,并将其返回给应用程序,然后应用程序把数据恢复到设备上。在恢复时,如果程序安装完毕且存在用户相关的备份数据,Android会自动执行恢复操作(应用程序也能够调用 BackupManager 的 requestRestore()方法发起恢复请求)。

1. 主要方法

BackupManager 提供的主要方法包括以下几种。

(1) dataChanged()和 dataChanged(String packageName)

为了获取一个备份,只需要创建一个 BackupManager 的实例,然后调用它的 dataChanged()方法。例如:

```
public void requestBackup() {
    BackupManager bm = new BackupManager(this);
    bm.dataChanged();
}
```

应用数据备份请求通过调用 BackupManager 对象的 dataChanged()方法发出。dataChanged()方法先调用 checkServiceBinder()方法(BackupManager 对象在发出发出备份和恢复的请求之前需要先调用 checkServiceBinder()方法实例化一个 BackupManagerService 服务的客户端代理对象,通过该对象向 BackupManagerService 服务发出备份或恢复请求。)获得 BackupManagerService 服务客户端代理对象后,接着调用服务端的 dataChanged()同名方法发出备份请求。

(2) requestRestore(RestoreObserver observer)

应用数据恢复请求通常由系统在应用安装时发现有要恢复的数据时自动触发,PackageManagerService 在应用安装成功后,如果发现新安装的工程文件包含 backupAgent 属性时自动触发。

当然用户也可以调用 BackupManager 对象的 requestRestore()方法来发出一个恢复请求。requestRestore()方法同样先调用 checkServiceBinder()方法获得 BackupManagerService 服务的一个客户端代理对象,然后调用服务端的 beginRestoreSession()方法实例化一个 ActiveRestoreSession 对象,返回客户端一个 IrestoreSession 接口对象,并根据 IrestoreSession 接口实例化一个客户端 RestoreSession 对象,通过 RestoreSession 对象使用 IrestoreSession 接口向服务端的 ActiveRestoreSession 对象的方法发送恢复请求。

并不是所有 Android 平台的设备都能支持数据备份。不过,就算是在不支持备份传输的设备上,程序仍然会正常运行,只是不能接收备份管理器的请求进行数据备份而已。

注意:备份服务并不是设计用于与其他客户端同步应用程序数据,或者用于在通常的应用程序生命周期中保存想访问的数据。不能够请求读、写备份数据,也不能够通过除备份管理器的 API 以外的方式访问备份数据。在备份过程中,只有备份管理器和备份传输器有权限访问被提交的数据。

2. 备份传输器

备份传输器是 Android 备份框架的客户端组件,它可由设备制造商和提供商定制。备份传输器可以因设备不同而不同,对于应用程序而言它是透明的。备份管理器的 API 将应用程序和实际备份传输器连接起来，通过一组固定的API与备份管理器进行通信，而不必关心底层的传输过程。

3. 检查数据版本

在把数据保存到云存储中时，备份管理器会自动包含应用程序的版本号，版本号是在AndroidManifest.xml文件的android:versionCode属性中定义的。在调用备份代理恢复数据之前，备份管理器会查询已安装程序的android:versionCode，并与记录在备份数据中的版本号相比较。如果备份数据的版本比设备上的要新，则意味着用户安装了旧版本的程序。这时备份管理器将停止恢复操作，onRestore()方法也不会被调用。

用android:restoreAnyVersion属性可以取代以上规则。此属性用true或false标明是否在恢复时忽略数据集的版本，默认值是false。如果将其设为true，备份管理器将忽略android:versionCode并且每次都会调用onRestore()方法。这时候可以在onRestore()里人工检查版本，并在版本冲突时采取必要的措施保证数据的兼容性。

为了便于在恢复数据时对版本号进行判断处理，onRestore()把备份数据的版本号作为appVersionCode参数和数据一起传入方法。而用PackageInfo.versionCode可以查询当前应用程序的版本号，例如：

```
PackageInfo info;
try {
   String name = getPackageName();
   info = getPackageManager().getPackageInfo(name,0);
} catch (NameNotFoundException nnfe) {
   info = null;
}

int version;
if (info != null) {
   version = info.versionCode;
}
```

然后，简单比较一下PackageInfo中的version和传入onRestore()的appVersionCode即可。

8.1.3 BackupAgent 备份代理

为了备份应用程序数据，需要实现一个备份代理。此备份代理将被备份管理器调用，用于提供所需备份的数据。当程序重装时，还要调用此代理来恢复数据。备份管理器处理所有与云存储之间的数据传输工作，备份代理则负责所有对设备上数据的处理。

BackupAgent是Android提供的两种备份代理方式之一。BackupAgent类提供了核心接口，程序通过这些接口与备份管理器进行通信。如果直接继承此类，必须重载onBackup()和onRestore()方法来处理数据的备份和恢复操作。

大多数应用程序应该不需要直接继承使用BackupAgent类，取而代之的是继承BackupAgentHelper类，并利用BackupAgentHelper内建的helper类自动备份和恢复文件。不过，如果需要实现以下目标，可以直接继承BackupAgent：

- 将数据格式版本化。例如需要在恢复数据时修正格式,可以建立一个备份代理,在数据恢复过程中如果发现当前版本和备份时的版本不一致,可以执行必要的兼容性修正工作。
- 不是备份整个文件,而是指定应当被复制的那一部分数据以及每一部分之后如何还原到设备上。这也有助于管理不同版本的数据,因为是把数据作为唯一 Entity 来读写,而不是读写整个文件。
- 备份数据库中的数据。如果用到 SQLite 数据库并且希望用户重装系统时能恢复其中数据,需要建立自定义的 BackupAgent。它在备份时读取库中数据,而在恢复时建表并插入数据。

通过继承 BackupAgent 创建备份代理时,必须实现以下回调方法。

1. onBackup()

备份管理器在程序请求备份后将调用本方法。在本方法中实现从设备读取应用程序数据,并把需备份的数据传递给备份管理器。

应用程序可以在任意时间调用 dataChanged()请求备份操作。这个方法提醒备份管理器,需要使用备份代理执行备份数据。接着,备份管理器会在某个合适的时刻回调备份代理的 onBackup()方法(注意,一次备份请求并不会在 onBackup()方法调用时立即返回结果)。一般来说,应当在每一次数据改变时请求一次备份(例如当用户更改了应用程序配置时)。如果在备份管理器从代理请求备份之前,连续调用 dataChanged(),代理依旧只调用一次 onBackup()。

当备份管理器调用 onBackup()方法时,它会传递以下三个参数。

(1) oldState

一个公开的、只读的 ParcelFileDescriptor 对象,指向由应用提供的上一次备份状态。这不是来自云储存的备份数据,而是本地数据的一个标记,是在上一次调用 onBackup()时备份的(由下面的 newState 定义,或者来自 onRestore())。由于 onBackup()不允许读取云储存上已有的备份数据,可以使用这个本地的标记来判断从上次备份以来数据是否发生改变。

(2) data

一个 BackupDataOutput 对象,用于传递备份数据到备份管理器。

(3) newState

一个公开的、可读可写的 ParcelFileDescriptor 对象,指向一个文件。在这个文件中,必须写入一个传递数据的标记(一个标记可以简单地使用文件最后一次修改的时间戳)。在下一次备份管理器调用 onBackup()方法时,这个对象会被当作 oldState 返回。如果没有向 newState 中写入备份数据,那么下一次备份管理器调用 onBackup()时,oldState 将会指向一个空文件。

下面的代码是使用 BackupAgent 实现备份代理时实现数据备份的示例,onBackup()方法的代码如下:

```
@Override
public void onBackup(ParcelFileDescriptor oldState, BackupDataOutput data,
        ParcelFileDescriptor newState) throws IOException {

```

```
05     synchronized (BackupRestoreActivity.sDataLock) {
06         RandomAccessFile file = new RandomAccessFile(mDataFile, "r");
07         mFilling = file.readInt();
08         mAddMayo = file.readBoolean();
09         mAddTomato = file.readBoolean();
10     }
11
12     boolean doBackup = (oldState == null);
13     if (!doBackup) {
14         doBackup = compareStateFile(oldState);
15     }
16
17     if (doBackup) {
18         ByteArrayOutputStream bufStream = new ByteArrayOutputStream();
19
20         DataOutputStream outWriter = new DataOutputStream(bufStream);
21         outWriter.writeInt(mFilling);
22         outWriter.writeBoolean(mAddMayo);
23         outWriter.writeBoolean(mAddTomato);
24
25         byte[] buffer = bufStream.toByteArray();
26         int len = buffer.length;
27         data.writeEntityHeader(APP_DATA_KEY, len);
28         data.writeEntityData(buffer, len);
29     }
30
31     writeStateFile(newState);
32 }
```

05～10 行通过 synchronized 实现用同一个内部锁的同步备份和恢复操作。12～15 行调用 compareStateFile()方法比较 oldState，检查自上次备份以来数据是否发生改变。从 oldState 读取信息的方式取决于当时写入的方式。最简单的记录文件状态的方式是写入文件的最后修改时间戳，compareStateFile()方法的代码如下：

```
boolean compareStateFile(ParcelFileDescriptor oldState) {
    FileInputStream instream = new FileInputStream(
            oldState.getFileDescriptor());
    DataInputStream in = new DataInputStream(instream);

    try {
        int stateVersion = in.readInt();
        if (stateVersion > AGENT_VERSION) {
            return true;
        }

        int lastFilling = in.readInt();
        boolean lastMayo = in.readBoolean();
        boolean lastTomato = in.readBoolean();
```

```
15
16          return (lastFilling != mFilling)
17                  || (lastTomato != mAddTomato)
18                  || (lastMayo != mAddMayo);
19      } catch (IOException e) {
20          return true;
21      }
22  }
```

02 行通过 FileInputStream 获取 oldState 输入流，07 行从 state 文件获取最后的修改时间戳。

onBackup()方法的 17～29 行是在与 oldState 比较后，如果数据发生了变化，则把当前数据写入 data，以便将其返回并上传到云存储中。

必须以 BackupDataOutput 中的 entity 方式写入每一块数据(程序负责把数据切分为多个 entity，当然也可以只用一个 entity)。一个 entity 是用一个唯一字符串键值标识的拼接二进制数据记录。因此，所备份的数据集其实是一组键值对。要在备份数据集中增加一个 entity，必须调用 writeEntityHeader()，传入代表写入数据的唯一字符串键值和数据大小。同时调用 writeEntityData()，传入存放数据的字节类型缓冲区，以及需从缓冲区写入的字节数(必须与传给 writeEntityHeader()的数据大小一致)。

无论是否执行备份，都要把当前数据的状态信息写入 ParcelFileDescriptor 对象 newState 指向的文件内。备份管理器会在本地保持此对象，以代表当前备份的数据。下次调用 onBackup()时，此对象作为 oldState 返回给应用程序，由此可以决定是否需要再做一次备份。如果不把当前数据的状态写入此文件，下次调用时 oldState 将返回空值。负责写状态信息的 writeStateFile()方法代码如下：

```
void writeStateFile(ParcelFileDescriptor stateFile) throws IOException {
    FileOutputStream outstream = new FileOutputStream(
            stateFile.getFileDescriptor());
    DataOutputStream out = new DataOutputStream(outstream);

    out.writeInt(AGENT_VERSION);
    out.writeInt(mFilling);
    out.writeBoolean(mAddMayo);
    out.writeBoolean(mAddTomato);
}
```

以下代码实现把文件最后修改时间戳作为当前数据的状态存入 newState：

```
FileOutputStream outstream = new FileOutputStream(
        newState.getFileDescriptor());
DataOutputStream out = new DataOutputStream(outstream);
long modified = mDataFile.lastModified();
out.writeLong(modified);
```

2. onRestore()

备份管理器在恢复数据时调用本方法(也可以主动请求恢复,但在用户重装应用程序时系统会自动执行数据恢复。),备份管理器调用本方法时将传入备份的数据,然后就可把数据恢复到设备上。

在应用程序正常的生命周期中,一般不需要请求还原。在应用程序安装时,系统会自动检查备份数据并执行还原。然而,如果有必要,可以调用 requestRestore()方法手动请求还原。在这种情况下,备份管理器调用 onRestore()方法实现,同时传递来自当前备份数据集的数据。

只用备份管理器能够调用 onRestore()。这发生在系统安装应用并发现已存在的备份数据的时候。然后,可以通过调用 requestRestore()请求还原操作。当备份管理器调用 onRestore()方法时,它传递以下三个参数。

(1) data。

一个 BackupDataInput 对象,它允许读取备份数据。

(2) appVersionCode。

一个整数,标记应用程序的 android:versionCode 清单的属性值,这个值是当前数据备份的时间。可以使用这个值来再次确认当前应用程序的版本,并判断数据格式是否兼容。

(3) newState。

一个公开的、可读可写的 ParcelFileDescriptor 对象,指向一个文件。在这个文件中,必须写入数据提供的最终备份状态。当下一次 onBackup()被调用,这个对象会作为 oldState 被返回。重复调用,必须写入同样的 newState 在 onBackup()的回调中。同时,这样可以确保传给 onBackup()的 oldState 对象是合法的,即使是在设备恢复后第一次调用 onBackup()。

下面的代码是使用 BackupAgent 实现备份代理时实现数据恢复的示例,onRestore()方法的代码如下:

```
@Override
public void onRestore(BackupDataInput data, int appVersionCode,
        ParcelFileDescriptor newState) throws IOException {
    while (data.readNextHeader()) {
        String key = data.getKey();
        int dataSize = data.getDataSize();

        if (APP_DATA_KEY.equals(key)) {
            byte[] dataBuf = new byte[dataSize];
            data.readEntityData(dataBuf, 0, dataSize);
            ByteArrayInputStream baStream = new ByteArrayInputStream(
                    dataBuf);
            DataInputStream in = new DataInputStream(baStream);

            mFilling = in.readInt();
            mAddMayo = in.readBoolean();
            mAddTomato = in.readBoolean();

            synchronized (BackupRestoreActivity.sDataLock) {
                RandomAccessFile file = new RandomAccessFile(mDataFile, "rw");
```

```
21                  file.setLength(0L);
22                  file.writeInt(mFilling);
23                  file.writeBoolean(mAddMayo);
24                  file.writeBoolean(mAddTomato);
25              }
26          } else {
27              data.skipEntityData();
28          }
29      }
30
31      writeStateFile(newState);
32  }
```

在 onRestore()实现中,应当像 04 行那样对 data 调用 readNextHeader()方法,以遍历数据集里所有的 entity。对其中每个 entity 需进行以下操作:

(1) 调用 getKey()获取 entity 的键值(如 05 行)。

(2) 将此 entity 键值和已知键值清单进行比较(如 08 行),这个清单应该已经在 BackupAgent 继承类中作为字符串常量(static final string)进行定义。一旦键值匹配其中一个键,就执行读取 entity 数据并保存到设备的语句。

① 调用 getDataSize()读取 entity 数据的大小并由此创建字节数组(如 06、09 行)。

② 调用 readEntityData(),传入字节数组作为获取数据的缓冲区,并指定起始位置和读取字节数(如 10 行)。

③ 字节数组将被填入数据,按需读取数据并写入设备即可。

(3) 把数据读出并写回设备以后,与上面备份数据的 onBackup()过程类似,调用 writeStateFile()方法把数据的状态写入 newState 参数。

8.1.4 BackupAgentHelper 备份代理

如果要备份整个文件(来自 SharedPreferences 或内部存储),应该用 BackupAgentHelper 创建备份代理来实现。BackupAgentHelper 类提供了 BackupAgent 类的易用性封装,它减少了需编写的代码数量(不必实现 onBackup()和 onRestore())。

BackupAgentHelper 的实现必须要使用一个或多个 Backup 助手。Backup 助手是一种专用组件,BackupAgentHelper 用它来对特定类型的数据执行备份和恢复操作。Android 框架目前提供两种 Backup 助手。

- SharedPreferencesBackupHelper:用于备份 SharedPreferences 文件。
- FileBackupHelper:用于备份来自内部存储的文件。

BackupAgentHelper 可包含多个 Backup 助手,但对于每种数据类型只需用到一个 Backup 助手。也就是说,即使存在多个 SharedPreferences 文件,也只需要一个 SharedPreferencesBackupHelper。

对于每个要加入 BackupAgentHelper 的 Backup 助手,必须在 onCreate()中执行以下步骤:

① 实例化所需的 Backup 助手。在其构造方法里必须指定需备份的文件。

② 调用 addHelper()。把 Backup 助手加入 BackupAgentHelper。

例如：

```
@Override
public void onCreate() {
   FileBackupHelper helper = new FileBackupHelper(this,
         TOP_SCORES, PLAYER_STATS);
   addHelper(FILE_HELPER_KEY, helper);
}
```

1. SharedPreferencesBackupHelper

SharedPreferencesBackupHelper 类包含了所有用于备份和还原一个 SharedPreferences 文件的代码。当备份管理器调用 onBackup()和 onRestore()时，BackupAgentHelper 调用备份助手来执行指定文件的备份和还原。

备份配置信息时，首先创建一个 SaredPreferencesBackupHelper。例如：

```
public class BlundellBackupAgent extends BackupAgentHelper {

   @Override
   public void onCreate() {
      super.onCreate();
      SharedPreferencesBackupHelper helper = new
            SharedPreferencesBackupHelper(
                  this,
                  PreferenceConstants.TUTORIAL_PREFERENCES);
      addHelper(PreferenceConstants.HELPER_KEY, helper);
   }
}
```

然后在 Activity 的 onCreate()生命周期方法中实例化 SharedPreferences 对象和 BackupManager 实例。例如：

```
@Override
public void onCreate(Bundle savedInstanceState) {
   super.onCreate(savedInstanceState);
   setContentView(R.layout.activity_backup_agent_helper);
   ...
   //Setup your shared preferences
   SharedPreferences sharedPreferences = getSharedPreferences(
         PreferenceConstants.TUTORIAL_PREFERENCES,
         MODE_PRIVATE);
   BackupManager backupManager = new BackupManager(this);
   CloudBackedSharedPreferences preferences = new
         CloudBackedSharedPreferences(sharedPreferences,
               backupManager);
}
```

最后,在 AndroidManifest.xml 的<application>标签内用 android:backupAgent 属性声明备份代理。例如:

```
<application android:label = "@string/app_name"
        android:backupAgent = ".backupagent.BlundellBackupAgent">
    ...
    <meta-data android:name = "com.google.android.backup.api_key"
    android:value = "AEdPqrEAAAAI679CqOut5LOOvfH60cqdyWz7TqhvQkkHdXN4ng"/>
    <activity ... >
        ...
    </activity>
</application>
```

注意:SharedPreferences 是线程安全的,所以可以从备份代理和其他 Activity 安全地读写共享配置文件。

2. FileBackupHelper

FileBackupHelper 用于备份一个文件。在实例化 FileBackupHelper 时,必须包含一个或多个保存于程序内部存储中的文件名称。下面是一个 FileBackupHelper 的示例:

```
public class FileHelperExampleAgent extends BackupAgentHelper {
    static final String FILE_HELPER_KEY = "the_file";

    @Override
    public void onCreate() {
        FileBackupHelper helper = new FileBackupHelper(this,
                BackupAgentActivity.DATA_FILE_NAME);
        addHelper(FILE_HELPER_KEY, helper);
    }

    @Override
    public void onBackup(ParcelFileDescriptor oldState,
            BackupDataOutput data,
            ParcelFileDescriptor newState) throws IOException {
        synchronized (BackupAgentActivity.sDataLock) {
            super.onBackup(oldState, data, newState);
        }
    }

    @Override
    public void onRestore(BackupDataInput data, int appVersionCode,
          ParcelFileDescriptor newState) throws IOException {
      synchronized (BackupAgentActivity.sDataLock) {
        super.onRestore(data, appVersionCode, newState);
      }
    }
}
```

这个 FileHelperExampleAgent 类包含所有必要的代码,以备份和还原保存到应用的内部存储文件。然而，在内部存储中读取和写入文件并不是线程安全的。为确保备份代理不在一个时间读取或写入文件，必须在每一次执行读或者写操作时使用同步的状态。例如，在任何的 Activity 内读取或写入文件时，需要使用一个对象为同步状态提供一个固有的锁：

```
//Object for intrinsic lock
static final Object sDataLock = new Object();
```

然后在每一次读取或写入文件时，使用这个锁创建一个同步的状态。例如，下面的同步状态用于将游戏中的最后一次得分写入一个文件中：

```
try {
   synchronized (MyActivity.sDataLock) {
       File dataFile = new File(getFilesDir(), TOP_SCORES);
       RandomAccessFile raFile = new RandomAccessFile(dataFile, "rw");
       raFile.writeInt(score);
   }
} catch (IOException e) {
   Log.e(TAG, "Unable to write to file");
}
```

应当使用同一个锁同步读取状态，接着在 BackupAgentHelper 类中必须重写 onBackup() 和 onRestore() 方法，使用同一个固有的锁同步备份和还原操作。例如，上面 FileHelperExampleAgent 这个例子需要以下方法：

```
@Override
public void onBackup(ParcelFileDescriptor oldState, BackupDataOutput data,
         ParcelFileDescriptor newState) throws IOException {
  //Hold the lock while the FileBackupHelper performs backup
   synchronized (MyActivity.sDataLock) {
       super.onBackup(oldState, data, newState);
   }
}

@Override
public void onRestore(BackupDataInput data, int appVersionCode,
       ParcelFileDescriptor newState) throws IOException {
   //Hold the lock while the FileBackupHelper restores the file
   synchronized (MyActivity.sDataLock) {
       super.onRestore(data, appVersionCode, newState);
   }
}
```

这样，就可以添加 FileBackupHelper 对象到 onCreate()方法中，并重写 onBackup()和 onRestore()方法以同步读写操作。

8.1.5 测试备份代理

实现了备份代理之后,可以使用 bmgr 命令工具测试备份和还原功能。

bmgr 是一个 Shell 工具,用来支持 Android 8 或更高版本 API 接口的 Android 设备的备份管理器进行交互。它提供了包括备份和恢复操作的命令,这样就不必反复擦除数据或采取类似的侵入性步骤来测试应用程序的备份代理功能了。这些命令均可在 adb shell 上运行。

1. 启动/禁用备份

通过 bmgr enabled 命令,可以看到备份管理器当前是否可以操作,例如:

```
adb shell bmgr enable < boolean
```

<boolean>是 true 或 false。这相当于在设备的主设置用户界面禁用或启用备份操作。

如果应用程序的备份代理从未被调用进行备份,那么验证操作系统认为它是否应该执行这样的操作,可能要用到这个命令。

注意:当备份功能被禁用了,当前的备份系统将会从备份存储里擦除整个有效的数据集。这就是说,当一个用户不希望他的数据被备份,备份管理器将不再备份。数据将不会被保存在设备中,也不可能有恢复操作,除非重新启用备份管理器。

2. 强制备份操作

通常,当应用程序的数据发生了变化,通过调用 dataChanged()方法来通知备份管理器。然后备份管理器将调用备份代理的 onBackup()接口。然而,可以通过运行 bmgr 命令来调用备份请求,例如:

```
adb shell bmgr backup < package >
```

<package>是计划进行备份的应用程序的正式包名。当执行了此备份命令,应用程序的备份代理将在未来一段时间内被调用来执行备份操作。虽然不能保证会备份,但是借助 bmgr 运行命令,仍可以强制所有还未进行的备份立即运行,例如:

```
adb shell bmgr run
```

立即执行的备份操作,包括自上次备份操作后调用 dataChanged()的所有应用程序和通过 bmgr backup 手动计划备份的任何应用程序。

3. 强制还原操作

与集中在一起、间或运行的备份操作不同,还原操作是立即执行的。备份管理器当前提供了两种类型的恢复操作。第一种是从从已备份的数据中恢复整个设备的数据。这通常是一个设备首次配置(复制用户先前设备的设置和其他保存的状态),并且仅由系统的操作执行。第二种还原操作是恢复单一的应用程序到“活动”的数据设置,即应用程序将舍弃当前的数据,从当前备份映像的最后一个运行良好的数据备份恢复。可以通过 requestRestore()方法调用第二个还原操作。备份管理器将调用备份代理的 OnRestore()接口。

在测试应用程序时,通过使用 bmgr 的恢复命令,可以立即调用应用程序的还原操作,例如：

```
adb shell bmgr restore <package>
```

<package>是想恢复的应用程序的正式包名，前提是这个应用程序已进行过备份、恢复操作。备份管理器将立即检查应用程序的备份代理，并调用它进行还原。即使应用程序当前没有运行，也可以进行。

4. 擦除数据

单个应用程序的数据可以从当前的数据集中擦除。当正在开发一个备份代理，如果错误导致写下无效的数据或保存状态信息，这个命令是非常有用的。可以用 bmgr 擦除命令来擦除应用程序的数据，例如：

```
adb shell bmgr wipe <package>
```

<package>是要删除数据的应用程序的正式包名。应用程序代理进程的下一次备份操作使这个应用程序看起来似乎之前从未备份过。

8.2 Google 云信息

Google 云信息(Google Cloud Messaging for Android，GCM)是一个能够帮助开发者从服务端发送数据到运行在 Android 手机程序的免费服务。这个服务提供了一个简单、轻量级的机制，使得服务端可以告诉移动端的程序与服务端建立直接的联系，来获取更新的程序或者用户的数据。

8.2.1 GCM 框架

GCM 是云到端消息框架(C2DM)的改进，实现了服务无配额限制、无须注册，并提供了一套更丰富的全新接口。Google GCM 服务器维护客户端和服务端之间的通信，负责处理消息队列和分发至运行在 Android 设备上的目标应用的各个方面。GCM 服务如今已成为 Google 其他众多接口的一部分，并由一个基于 Google API 控制台的项目所管理。与 Google 其他接口不同，GCM 服务没有配额限制，所以无论有多少消息、多少设备使用这项服务，都是完全免费的。

1. 主要特点

GCM 服务的主要特点包括：

- 允许第三方应用服务器发送消息到 Android 应用。
- 不能保证消息的发送和消息的顺序。
- 不需要 Android 设备上的应用一直运行，在收到消息时，只要在应用的 AndroidManifest.xml 中设置了适当的 Permission 和 BroadcastReceiver，系统就可以通过 Intent 来唤醒应用。
- 不提供任何内置的用户界面或其他对消息数据的处理方法。GCM 只是简单地把原始数据传递给 Android 应用程序,而程序会负责如何处理消息数据。

- 需要可以运行 Android API 8 或者更高版本而且要装有 Google Play Store 的设备。
- 使用现有的 Google 服务连接。对于 API 11 以前的设备,需要用户在移动设备上设置 Google 账户。运行 Android API 14 或更高版本的设备是不需要 Google 账户的。

2. GCM 的核心组件

GCM 工作的核心包括两个方面。

(1) 组件

GCM 包含的组件包括以下几种。

- 移动设备:运行使用 GCM 的 Android 应用的设备。
- 第三方应用服务:开发人员用以作为实现 GCM 一部分的应用服务器。第三方服务器通过 GCM 服务器给设备上的应用发送数据。
- GCM 服务器:Google 服务器从第三方服务器上获取数据,并把它们发送到移动设备上。

(2) 认证

用在不同阶段来确认各方都已经被认证的 ID 和令牌,确保消息能发到正确的地方。

- Sender ID:从 API 控制台获取的项目 ID。Sender ID 被用在注册阶段,用来确认 Android 应用是否已经被允许发消息给设备。
- Application ID:Android 程序注册应用程序的 ID,用来获得消息。Android 程序是通过 AndroidManifest. xml 中的包名来区分的。这确保该消息是针对正确的 Android 应用程序。
- Registration ID:由 GCM 服务器发送给 Android 应用,允许应用接收消息。一旦 Android 应用拥有了 Registered ID,就把它发送给第三方应用服务器,服务器用它来确认哪些设备已经注册了并且正准备接收消息。换句话说,Registered ID 被绑定于运行在特殊设备上的特殊应用上。
- Google 用户账号:为了 GCM 的工作,移动设备必须至少包含一个 Google 账户。
- 消息发送者验证令牌:保存在第三方应用服务器上的 API key,允许服务器访问 Google 服务。API key 存在于消息发送请求的头部信息里。

8.2.2 GCM 的事件序列

1. 客户端事件序列

以下是移动设备上的应用注册并接收消息时发生的事件序列:

(1) 第一次应用使用消息服务时,给 GCM 发送一个注册 Intent。该 Intent 的动作为 com. google. android. c2dm. intent. REGISTER,并包括发送者 ID 和 Android 应用程序的 ID。

(2) 如果注册成功,GCM 会广播一个动作为 com. google. android. c2dm. intent. REGISTER 的 Intent,返回给应用一个注册 ID。应用会在以后用到这个 ID(例如,会在 onCreate()方法里校验是否已经注册)。

注意:Google 可能会定期刷新注册 ID,所以在设计应用的时候,要认识到 com. google. android. c2dm. intent. REGISTER 可能会被多次调用。Android 应用程序需要能够做出相应的反应。

(3) 完成注册。Android 应用程序把注册 ID 发送到应用程序服务器。服务器把注册

ID 存储在数据库中。注册 ID 一直持续到 Android 的应用程序显式地注销它，或者 Google 对它进行刷新。

(4) 发送消息。应用程序服务器将消息发送到一个 Android 应用程序时，必须满足如下条件：

- 应用必须要有一个注册 ID，这样允许它在一个特定的设备上接收消息。
- 第三方服务器已经存储了注册 ID。
- 开发者必须已经在应用服务器上为应用准备好 API key。

(5) 接收消息。手机收到消息后，从消息中提取键值对，系统使用 com. google. android. c2dm. intent. RECEIVE Intent 把键值对传给目标程序，目标程序从 RECEIVE Intent 中根据 Key 取得数据并处理数据。

2. 服务端时间序列

以下是应用服务器发消息时发生的事件序列：

(1) 应用服务器给 GCM 发消息。

(2) Google 会给消息排序并存储它们(当设备不在线的时候)。

(3) 当设备在线的时候，Google 会把消息发送给它们。

(4) 在设备上，系统会使用合适的权限通过 Intent 广播把消息广播给具体的应用。消息会唤醒接收消息的应用。

(5) 接收并处理消息。设备上的应用接收消息时，接收送到的消息并从消息中提取键值对，通过 com. google. android. c2dm. intent. RECEIVE 把键值对信息发送给应用，应用通过 com. google. android. c2dm. intent. RECEIVE 提取数据并加以处理。

8.2.3　开发云信息服务

1. 创建 Google API 工程

创建一个 Google API 工程的步骤如下：

(1) 打开 Google APIs Console 页面(https://code. google. com/apis/console)。

(2) 如果未曾创建一个 API 工程，本页面会显示创建 API 工程的提示。

如果已经有工程了，看到的第一页将是 Dashboard 页面，可以通过打开工程下拉菜单的 Other projects→Create 来创建一个新的工程，如图 8-1 所示。

(3) 创建工程后再单击它，浏览器会变换到类似如下的链接：

```
https://code.google.com/apis/console/b/0/?noredirect#project:670597347343
```

#/project 的值(本例 670597347343)是工程的 ID，会在后面作为 GCM 传送 ID。

2. 激活 GCM 服务

激活 GCM 服务的步骤如下：

(1) 在 Google APIs Console 主页选择 Services，打开 Google Cloud Messaging for Android 开关。如图 8-2 所示。

(2) 在 Service 页的 Terms 中选择各选项。

3. 获取 API Key

获取 API Key 的步骤如下：

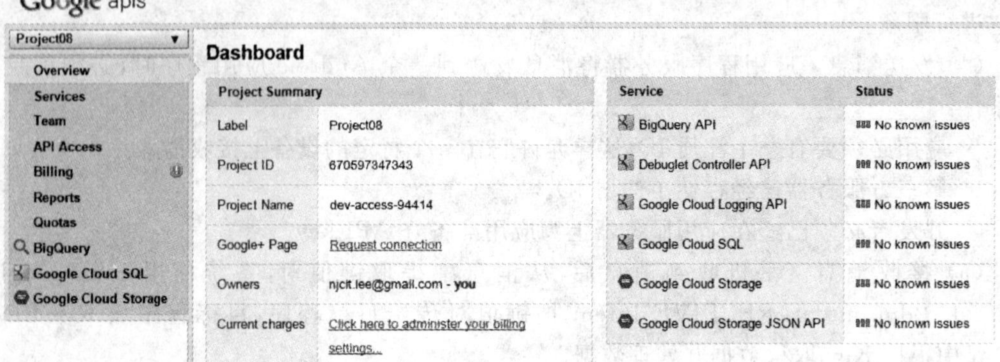

图 8-1 Google Dashboard 面板

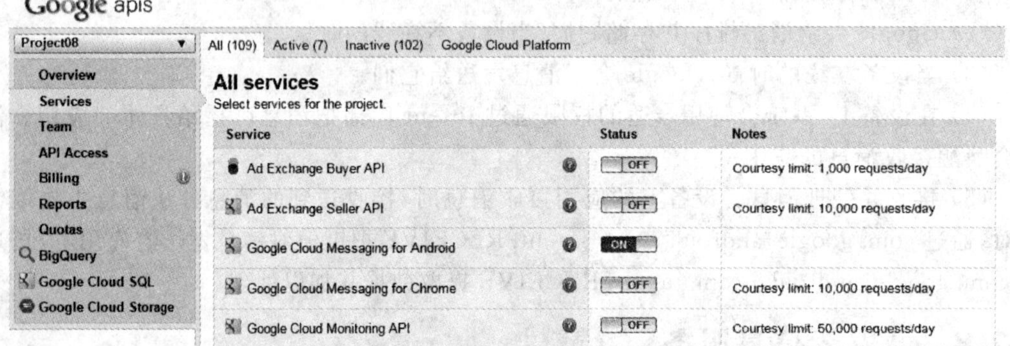

图 8-2 开启 Google Cloud Messaging for Android

(1) 在 Google APIs Console 主页选择左侧的 API Access。

(2) 在 API Access 页面,单击 Create new Server key 按钮,显示 Configure Server Key 对话框。

(3) 在"Accept requests from these server IP addresses:"输入框中输入服务器 IP 地址(可选,不输也可以),单击"创建"按钮,显示如图 8-3 所示的创建结果(本例 API key 结果为 AIzaSyDruFSRm81xHDeCNyFAH1KWhO3Gf381hMs)。

4. 创建客户端应用

在创建 Android 客户端应用之前,需要下载帮助库。方法是在 SDK Manager 的 Extras 中选中并下载 Google Cloud Messaging for Android Library。下载完成后,在 Android SDK 路径下的/extras/google/文件夹中将包含一个 GCM 文件夹,含如下子目录:gcm-client、gcm-server、samples/gcm-demo-client、samples/gcm-demo-server 和 samples/gcm-demo-appengine。

然后按照如下步骤创建 Android 应用。

(1) 创建 Android 应用,并将 gcm-client/dist 中的 gcm.jar 文件添加到应用的 Java Build Path 中。

(2) 修改应用的 AndroidManifest.xml 文件。包括以下方面:

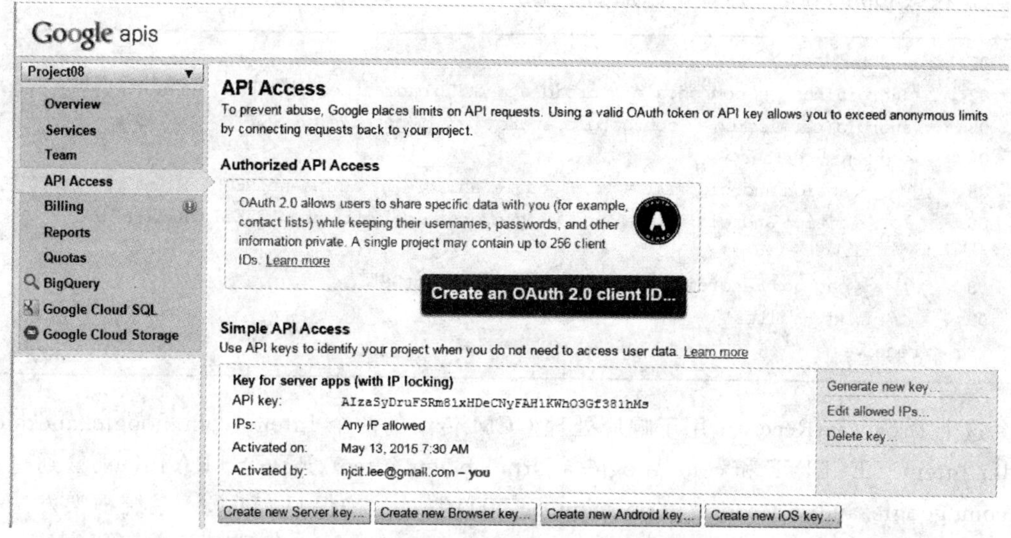

图 8-3　Server Key 显示界面

① 设置 Android SDK 最低版本号为 8，声明并使用自定义权限，保证只有此应用程序才能接收 GCM 信息，例如：

```
<permission
        android:name = "cn.liwy.project08.permission.C2D_MESSAGE"
        android:protectionLevel = "signature" />

<uses - permission
        android:name = "cn.liwy.project08.permission.C2D_MESSAGE" />
```

此 permission 必须命名为 app_package. permission. C2D_MESSAGE(app_package 是应用程序的包名，定义在 AndroidManifest. xml 标签中)，以防止其他程序注册和接收这个程序的消息。

② 添加如下 permission：

```
<uses - permission android:name =
        "com.google.android.c2dm.permission.RECEIVE" />
<uses - permission android:name = "android.permission.INTERNET" />
<uses - permission android:name = "android.permission.GET_ACCOUNTS" />
<uses - permission android:name = "android.permission.WAKE_LOCK" />
```

这些权限的含义是：

- 程序拥有注册和接收消息的权限。
- 程序拥有联网的权限。
- GCM 需要 Google 账户(设备版本低于 4.0 时)。
- 程序可以保证处理器在睡眠时被唤醒。

③ 在<application>标签中添加如下 BroadcastReceiver 声明：

```
<receiver
    android:name = "com.google.android.gcm.GCMBroadcastReceiver"
    android:permission = "com.google.android.c2dm.permission.SEND" >
    <intent - filter>
        <action android:name = "com.google.android.c2dm.intent.RECEIVE" />
        <action android:name = "com.google.android.c2dm.intent.REGISTRATION" />

        <category android:name = "cn.njcit.project08" />
    </intent - filter>
</receiver>
```

这个 BroadcastReceiver 用于响应处理 GCM 传给的两个 Intent(com. google. android. c2dm. intent. RECEIVE 和 com. google. android. c2dm. intent. REGISTRATION)。通过设定 com. google. android. c2dm. permission. SEND 的权限，确保只有 GCM 系统框架发送的 Intent 才发送给该广播。注意，在 category 标签里的 android:name 必须是应用的包名。

以下是 GCMBroadcastReceiver 的示例：

```
class GCMBroadcastReceiver extends BroadcastReceiver {
    @Override
    public void onReceive(Context context, Intent intent) {
        String regId = intent
                .getStringExtra(Constants.FIELD_REGISTRATION_ID);
        sendIdToServer(regId);
    }
}
```

(3) 编写 app_package. GCMIntentService 类，该类会被 GCMBroadcastReceiver 调用，并且必须是 com. google. android. gcm. GCMBaseIntentService 的子类，必须包含一个公共构造方法，而且应当被命名为 app_package. GCMIntentService(除非使用 GCMBroadcastReceiver 的子类重写方法来命名服务)。

例如：

```
public class GCMIntentService extends GCMBaseIntentService {
    @Override
    protected void onRegistered(Context context, String regId) {
        Intent intent = new Intent(Constants.ACTION_ON_REGISTERED);
        intent.putExtra(Constants.FIELD_REGISTRATION_ID, regId);
        context.sendBroadcast(intent);
    }

    @Override
    protected void onUnregistered(Context context, String regId) {
        Log.i(Constants.TAG, "onUnregistered: " + regId);
    }

```

```
    @Override
    protected void onMessage(Context context, Intent intent) {
        String msg = intent.getStringExtra(Constants.FIELD_MESSAGE);

        NotificationManager manager = (NotificationManager) context
                .getSystemService(Context.NOTIFICATION_SERVICE);
        Notification notification = prepareNotification(context, msg);
        manager.notify(R.id.notification_id, notification);
    }

    private Notification prepareNotification(Context context, String msg) {
        long when = System.currentTimeMillis();
        Notification notification = new Notification(
                R.drawable.ic_stat_cloud,msg, when);
        notification.flags |= Notification.FLAG_AUTO_CANCEL;

        Intent intent = new Intent(context, MessageActivity.class);
        //Set a unique data uri for each notification to make sure the activity
        //gets updated
        intent.setData(Uri.parse(msg));
        intent.putExtra(Constants.FIELD_MESSAGE, msg);
        intent.addFlags(Intent.FLAG_ACTIVITY_NEW_TASK
                | Intent.FLAG_ACTIVITY_CLEAR_TASK);
        PendingIntent pendingIntent = PendingIntent.getActivity(context, 0,
                intent, 0);
        String title = context.getString(R.string.app_name);
        notification.setLatestEventInfo(context, title, msg, pendingIntent);

        return notification;
    }

    @Override
    protected void onError(Context context, String errorId) {
        Toast.makeText(context, errorId, Toast.LENGTH_LONG).show();
    }
}
```

- onRegistered(Context context, String regId)：接收到一个注册过的 Intent 后被调用，作为一个参数传递 GCM 分配的注册 ID 给设备中相应的应用程序。典型地，应当发送 regid 给服务端以使它能发送消息给设备。
- onUnregistered(Context context, String regId)：当设备从 GCM 注销注册后被调用，典型地，应当发送 regid 给服务端以注销设备。
- onMessage(Context context, Intent intent)：当服务端发送消息给 GCM 调用，GCM 传送给设备。如果消息拥有 payload，它的 Content 可以作为 Intent 的 extras 来获得。
- onError(Context context, String errorId)：当设备试图要注册或注销，但 GCM 返回一个错误时调用。

- onRecoverableError(Context context, String errorId)：(可选)当设备试图要注册或注销，但是 GCM 服务不可获得，GCM 库会使用指数备份尝试重复操作，除非此方法被重写且返回 false。此方法是可选方法，只有当想显示消息给用户或取消重复尝试才应当被重写。例如：

```
@Override
protected boolean onRecoverableError(Context context, String errorId) {
    Log.e(TAG, "Received recoverable error: " + errorId);
    return super.onRecoverableError(context, errorId);
}
```

上面的这些方法运行在 IntentService 的线程，可以任意进行网络调用，不会阻塞 UI 线程。

最后在 AndroidManifest.xml 中添加类似如下的 IntentService 声明：

```
<service android:name = "cn.liwy.project08.GCMIntentService" />
```

(4) 编写应用的主 Activity。

应用的主 Activity 的 onCreate()方法一般为：

```
@Override
public void onCreate(Bundle savedInstanceState) {
    super.onCreate(savedInstanceState);
    ...

    mGCMReceiver = new GCMBroadcastReceiver();
    mOnRegisteredFilter = new IntentFilter();
    mOnRegisteredFilter.addAction(Constants.ACTION_ON_REGISTERED);

    if (Constants.SENDER_ID == null) {
        mStatus.setText("Missing SENDER_ID");
        return;
    }
    if (Constants.SERVER_URL == null) {
        mStatus.setText("Missing SERVER_URL");
        return;
    }

    GCMRegistrar.checkDevice(this);
    GCMRegistrar.checkManifest(this);
    final String regId = GCMRegistrar.getRegistrationId(this);
    if (!regId.equals("")) {
        sendIdToServer(regId);
    } else {
        GCMRegistrar.register(this, Constants.SENDER_ID);
    }
}
```

checkDevice()方法检查 GCM 的设备系统版本及是否装有 Google Service Frame(GCM 服务需要安装有谷歌服务包,且系统版本 2.2 及以上的才支持),如果不支持则抛出异常。代码如下:

```
private boolean checkDevice() {
    int resultCode = GooglePlayServicesUtil
            .isGooglePlayServicesAvailable(this);
    if (resultCode != ConnectionResult.SUCCESS) {
        if (GooglePlayServicesUtil.isUserRecoverableError(resultCode)) {
            GooglePlayServicesUtil.getErrorDialog(resultCode, this,
                    PLAY_SERVICES_RESOLUTION_REQUEST).show();
        } else {
            Log.i(TAG, "This device is not supported.");
            finish();
        }
        return false;
    }
    return true;
}
```

类似地,checkManifest()验证应用的 AndroidMenifest.xml 是否包含了在编写 Android 应用所要求的条件。一旦完成合理的验证,设备调用 GCMRegsistrar.register()来注册设备,传递注册时获得的 SENDER_ID。可以先调用 CMRegistrar.getRegistrationId()来确认设备是否已经注册。然后当注册的 Intent 到来后,GCMRegistrar 单独跟踪注册 ID。

8.2.4 Google App Engine

Google App Engine(简称 GAE),是 Google 在 2008 年 Campfire One 上推出的支持 Python、Java、Go 和 PHP 语言的云引擎。GAE 是一种简化创建、运行和构建伸缩性 Web 应用的工具,用户可以在 Google 的基础架构上构建 Web 应用并将其部署到 Google 基础设施之上。GAE 易于构建和维护,并可根据访问量和数据存储需要的增长轻松扩展。GAE 对全球开发者免费开放使用,可以充分利用 Google 提供的免费空间、免费数据库、免费二级域名等来展示自己的应用程序,提供给全球的用户下载和使用。

1. 注册 GAE

注册 GAE 的步骤如下:

(1) 注册 Google Gmail 邮箱(网址为 https://accounts.google.com/signup)并完成验证。

(2) 登录 Google App Engine(网址为 https://appengine.google.com/),显示如图 8-4 所示。

(3) 单击 Create Application 按钮,显示如图 8-5 所示。

(4) 在 Application Idenifier 的 appspot.com 之前的输入框中输入要上传的 appid,如果已经有人使用该 appid 创建过了,则提示失败。在 Application Title 部分的输入框中填入应用标题部分(可随意填入),其他选项缺省设置。单击底部的 Create Application,显示如图 8-6 所示,表明 appid 建立成功。

Google app engine njcit.lee@gmail.com

Welcome to Google App Engine

Before getting started, you want to learn more about developing and deploying applications.
Learn more about Google App Engine by reading the Getting Started Guide, the FAQ, or the Developer's Guide.

Create Application

图 8-4 Google App Engine 页面

Google app engine njcit.lee@gmail.com | My Account | Help | Sign out

Create an Application

You have 24 applications remaining.

Application Identifier:

[ ] .appspot.com

All Google account names and certain offensive or trademarked names may not be used as Application Identifiers.
You can map this application to your own domain later. Learn more

Application Title:

[ ]

Displayed when users access your application.

Authentication Options (Advanced): Learn more

Google App Engine provides an API for authenticating your users, including **Google Accounts**, **Google Apps**, and **OpenID**. If you choose to use this feature for some parts of your site, you'll need to specify now what type of users can sign in to your application:

- **Open to all Google Accounts users (default)**
  If your application uses authentication, anyone with a valid Google Account may sign in.
- **Restricted to the following Google Apps domain:**
  [ ]
  e.g. foo.com
  If your application uses authentication, **only members of this Google Apps domain may sign in.** If your organization uses Google Apps, use this option to create an application (e.g. an HR tracking tool) that is only accessible to accounts on your Google Apps domain. This option cannot be changed once it has been set.
- **(Experimental) Open to all users with an OpenID Provider**
  If your application uses authentication, anyone who has an account with an OpenID Provider may sign in.

Create Application | Cancel

图 8-5 创建 Google App Engine 应用页面

Google app engine njcit.lee@gmail.com | My Account | Help | Sign out

Application Registered Successfully

The application will use **njcit2015** as an identifier. This identifier belongs in your application's configuration as well. Note that this identifier cannot be changed. Learn more

The application uses the **High Replication** storage scheme. Learn more

If you use Google authentication for your application, **Contact App** will be displayed on Sign In pages when users access your application.

Choose an option below:

- View the dashboard for Contact App.
- Use appcfg to upload and deploy your application code.
- Add administrators to collaborate on this application.

© 2015 Google | Terms of Service | Privacy Policy | Blog | Discussion Forums | Project | Docs

图 8-6 Google App Engine 应用创建成功的页面

(5) 单击 dashboard 链接，显示如图 8-7 所示。

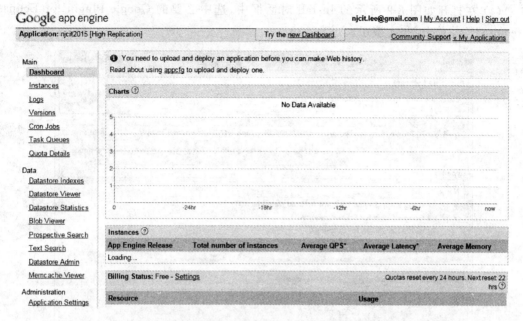

图 8-7　Google App Engine 应用于 Dashboard 面板

后续可以通过 Eclipse 上传应用项目。

2. 安装 GPE 插件

Google Plungin for Eclipse(简称 GPE)是 Google 推出的一款 Eclipse 插件，使用该插件，可以从 Eclipse 中创建、测试和上传 App Engine 应用程序，它同时支持 Google Web Toolkit(GWT)开发。据 Google Web Toolkit 的官方博客介绍，GPE 其实包括了一系列软件开发工具，用于帮助 Java 开发者快速设计、构建、优化及部署那些基于云的应用程序——使用 GWT、Speed Tracer、App Engine 及其他 Google 云服务。

安装 GPE 插件的步骤是：

(1) 启动 Eclipse，执行 Help → Install New Software...命令，打开 Install 对话框。

(2) 在 Install 对话框中单击 Add 按钮，打开 Add Repository 对话框，如图 8-8 所示。

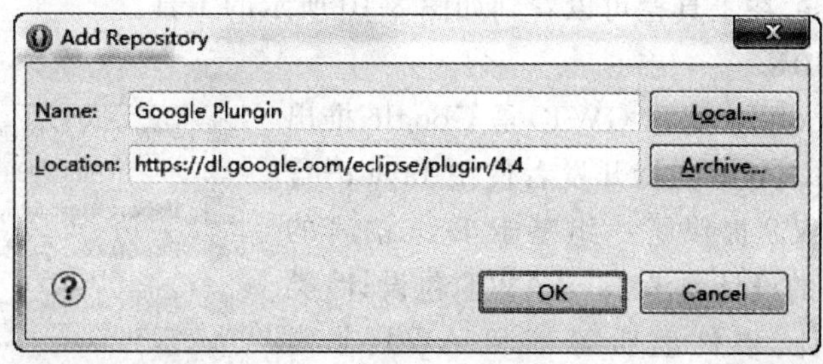

图 8-8　安装 GPE 插件

(3) 在 Location 输入框中输入 https://dl.google.com/eclipse/plugin/4.4(4.4 是根据 Eclipse 的版本号选择的，不同版本 Eclipse 的插件下载地址不同，Luna 版本号是 4.4，Kepler 是 4.3，Juno 是 4.2，访问 https://developers.google.com/eclipse/docs/download

可以查看更多不同版本的下载地址),单击 OK 按钮。

(4) 在打开如图 8-9 所示的 Install 对话框中,选中必要的 Google Plugin for Eclipse (required),并单击 Next 按钮。

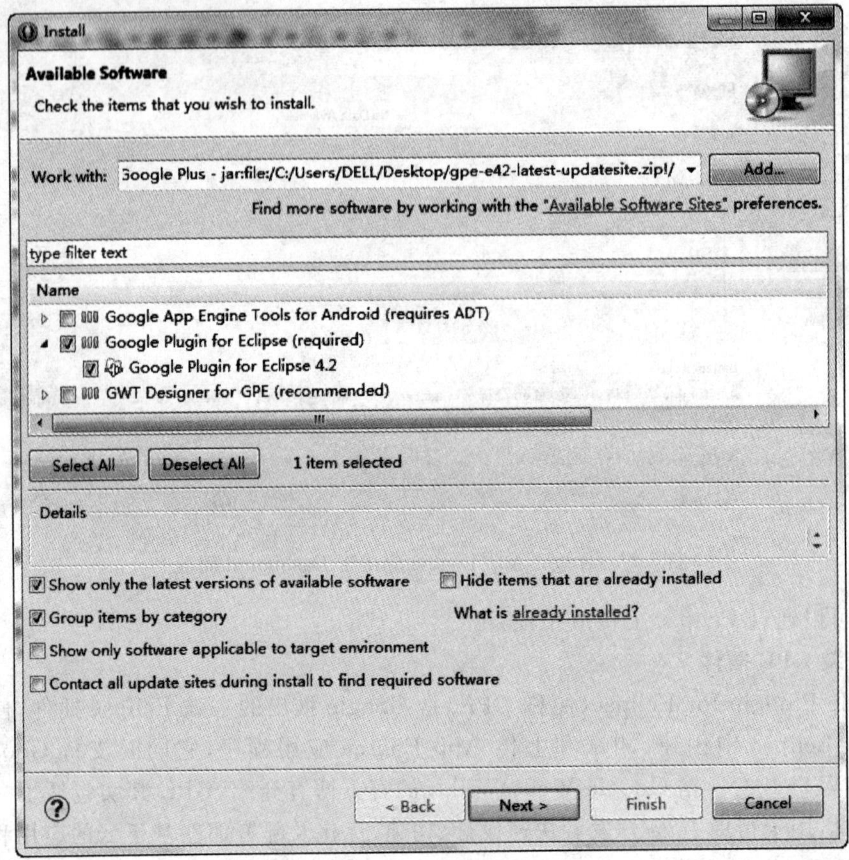

图 8-9 GPE 插件安装详细信息

(5) 阅读所要安装内容的细节及相关条款,选中 I accept the terms of the license agreements 单选按钮,从而确认进入下一步安装。

(6) 重启 Eclipse,在工具栏可以看到如图 8-10 所示的工具。

3. 安装 GWT SDK

Google Web Toolkit(简称 GWT)是 Google 推出的 Ajax 应用开发包,GWT 支持开发者使用 Java 语言开发 Ajax 应用。GWT 提供了一组基于 Java 语言的开发包,这个开发包的设计参考 Java AWT 包设计,类命名规则、接口设计、事件监听等都和 AWT 非常类似。

图 8-10 GPE 工具栏

安装 GWT SDK 的步骤是:

(1) 下载 GWT SDK(下载地址为 http://www.gwtproject.org/download.html)。

(2) 启动 Eclipse，执行 Window→Preferences 命令，打开 Preferences 对话框。

(3) 在 Preferences 对话框中展开左侧的 Google/Web Toolkit，单击 Add 按钮，在打开的 Add GoogleWeb Toolkit SDK 对话框中单击 Browse 按钮，选择下载的 GWT SDK，如图 8-11 所示。

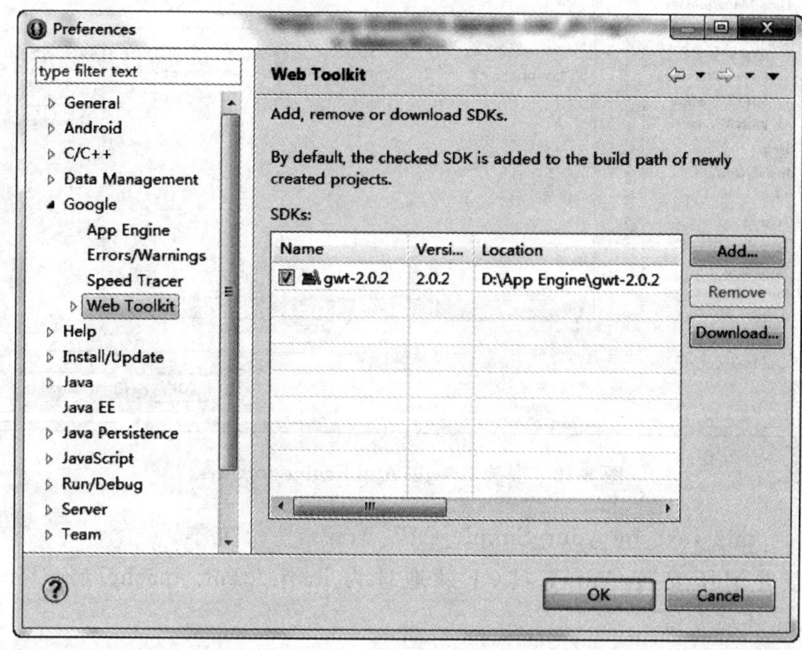

图 8-11　安装 GWT SDK

4. 安装 Google App Engine for Java SDK

Google App Engine for Java 为企业 Java 开发提供了一个端到端的解决方案：一个易于使用的基于浏览器的 Ajax GUI、Eclipse 工具支持以及后端的 Google App Engine。易于使用和工具支持是 Google App Engine for Java 优于其他云计算解决方案的两大优势。

安装 Google App Engine for Java 的步骤是：

(1) 下载 Google App Engine for Java（下载地址为 https://cloud.google.com/appengine/downloads）。

(2) 启动 Eclipse，执行 Window → Preferences 命令，打开 Preferences 对话框。

(3) 在 Preferences 对话框中展开左侧的 Google/App Engine，单击 Add 按钮，在打开的 Add App Engine SDK 对话框中单击 Browse 按钮，选择下载的 Google App Engine for Java SDK，如图 8-12 所示。

8.2.5　创建服务端应用

本小节通过 App Engine for Java 来构建云信息服务端。步骤如下：

(1) 在安装完 Google Cloud Messaging for Android Library 后，Android SDK 目录下有一个 extras/google/目录，进入 samples/gcm-demo-appengine 目录。

(2) 打开 gcm-demo-appengine/src/com/google/android/gcm/demo/server/目录下的 ApiKeyInitializer.java 文件，用 8.2.3 小节获取的 API Key 替换掉 entity.setProperty()方

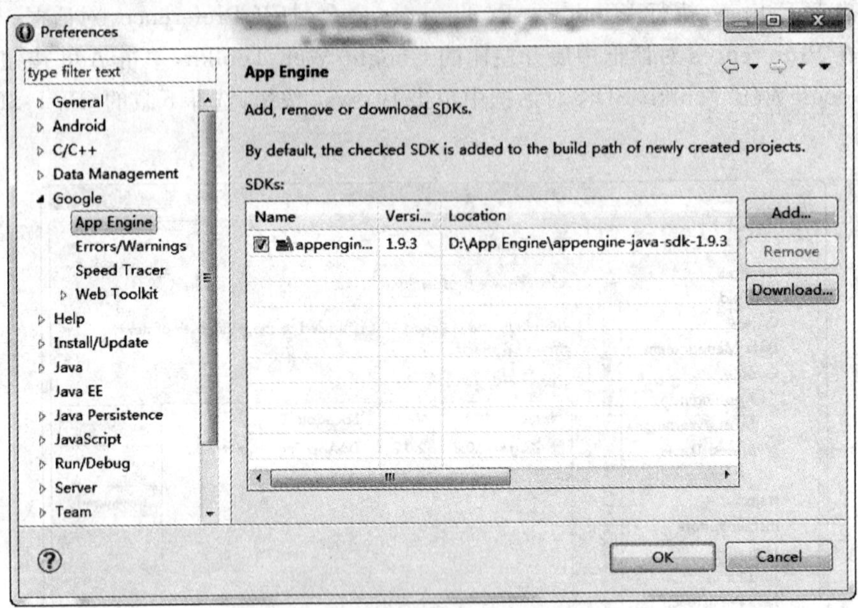

图 8-12　安装 Google App Engine for Java

法里的 replace_this_text_by_your_Simple_API_Access_key 文本。

(3) 下载并解压安装 Ant 工具(下载地址为 http://ant.apache.org/bindownload.cgi),配置 Ant 环境变量。

(4) 打开命令行窗口,进入 gcm-demo-appengine 所在的文件夹,运行如下命令:

```
ant -Dsdk.dir=D:\\App runserver -Dserver.host=192.168.0.105
```

其中,-Dsdk.dir 的值为 appengine-java-sdk 所在的文件夹,-Dserver.host 是服务器的 IP 地址。

如果相关配置和命令执行正确,在命令行窗口显示类似如下的信息:

```
Buildfile: C:\Users\DELL\workspace2015\gcm-demo-appengine\build.xml
init:
copyjars:
compile:
datanucleusenhance:
  [enhance] DataNucleus Enhancer (version 1.1.4) : Enhancement of classes
  [enhance] DataNucleus Enhancer completed with success for 0 classes. Timings :
input = 91 ms, enhance = 0 ms, total = 91 ms. Consult the log for full details
  [enhance] DataNucleus Enhancer completed and no classes were enhanced. Consult
the log for full details
...
```

(5) 打开浏览器输入 http://192.168.0.105:8080/,如果显示如图 8-13 所示的界面信息,说明服务端部署成功。

图 8-13 服务端界面

(6) 在设备上添加 Google 账号(在 Android 4.0 及以上版本中,添加账号是可选的,也就是可以省掉添加账号这一环节),添加账号的位置如图 8-14 所示。

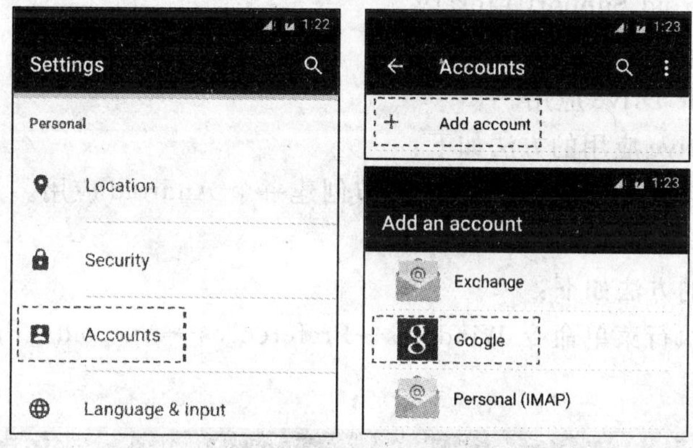

图 8-14 添加 Google 账户

(7) 启动 Android 客户端应用,刷新浏览器,此时将显示设备已经成功注册,客户端和服务端可以开始通信。

8.3 Google Drive

Google Drive 是 Google 的一个在线同步存储服务,并提供了如下核心功能:

- 结合 Google Docs 的在线文件预览与编辑功能。
- 版本控制功能,一个文件可以保存 30 天内的各个修订版本,并且仅仅只有当前最新的文件占用空间。
- 强大和完善的共享、协作功能,可以创建简单的共享链接供所有人浏览、评论,还可以很轻松地创建共享文件和文件夹,实现团队或朋友之间的文档共享,共同维护和编辑文档。
- 与 Gmail、Google+ Photos 的整合,提高了用户体验。

说明:当用户将资料上传或提交到 Google Drive 上时,用户即授权 Google(以及与 Google 合作的公司)在世界范围内使用、托管、存储、复制、修改、再创作(如翻译、改写或其他能够使用户的内容与 Google Drive 的服务更匹配的改变)、传播、出版、公共场合演示或分发资料的权利。用户授予的这些权利仅会用在 Google Drive 的运行维护、产品升级和服务提升等有限的用途上,即便用户停止使用 Google Drive,该服务条款仍然生效。

8.3.1 获取 Google Drive API Key

获取 Google Drive 应用的基本步骤如下。

(1) 安装 Google Play services SDK。

安装 Google Play services SDK 的方法如下:

在 Eclipse 中,执行菜单命令 Windows→Android SDK Manager,进入 Android SDK Manager 对话框,安装如下 3 个项目。

- Android 版本号/Google APIs(可选)
- Extras/Android Support Library
- Extras/Google Play services

(2) 注册 Google Drive 应用。

注册 Google Drive 应用的方法如下:

如果没有创建应用项目,参照 8.2.3 小节创建一个 Android 应用。

(3) 获取认证指纹。

获取认证指纹的方法如下:

在 Eclipse 中,执行菜单命令 Windows→Preferences→Android→Build,打开如图 8-15 所示对话框。

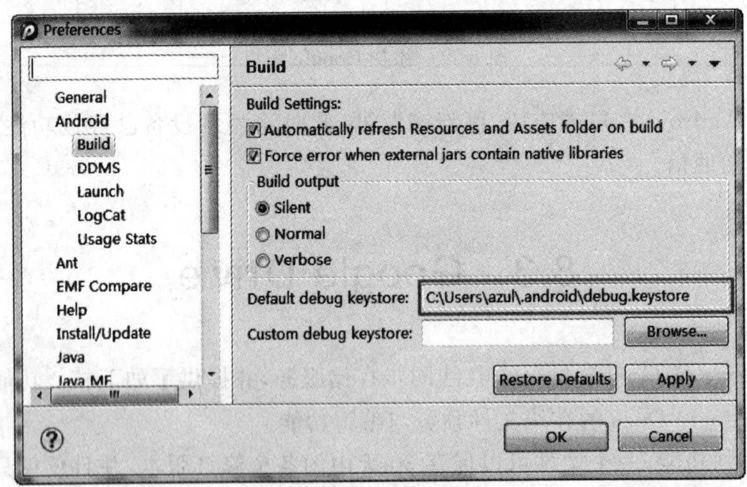

图 8-15 Preferences 对话框

其中的 Default debug keystore 即为 debug keystore 的位置。

打开命令行窗口,执行如下命令:

```
keytool -list -alias androiddebugkey -keystore "C:\\Users\\DELL\\.android\\debug.keystore" -storepass android -keypass android (注意:双引号内不得含空格)
```

keytool 为 jdk 中的命令，如果没有配置 JDK 环境变量，则进入 jdk/bin 目录中执行该命令。显示结果如图 8-16 所示。

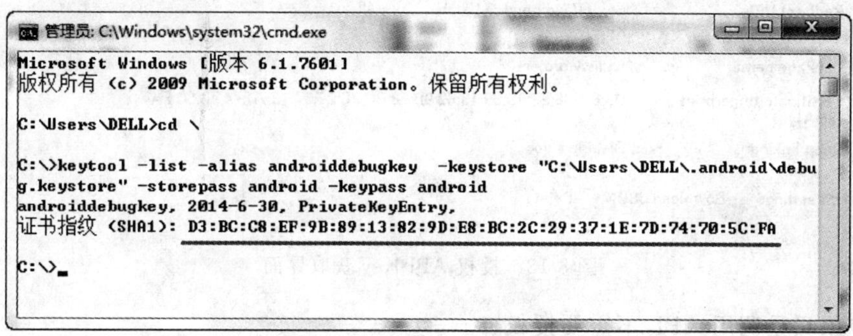

图 8-16 获取认证指纹

其中 SHA1 那一行就包含了证书的 SHA-1 认证指纹，它是二十段用冒号割开的数字段，每段是两个十六进制的数。

(4) 获取 API Key。

获取 API Key 的方法如下：

在 Web 浏览器中打开网址 https://code.google.com/apis/console，单击左侧 APIs & auth 下的 APIs，选择右侧的 Enabled APIs 标签页，查看 Drive API 是否被激活，如图 8-17 所示。

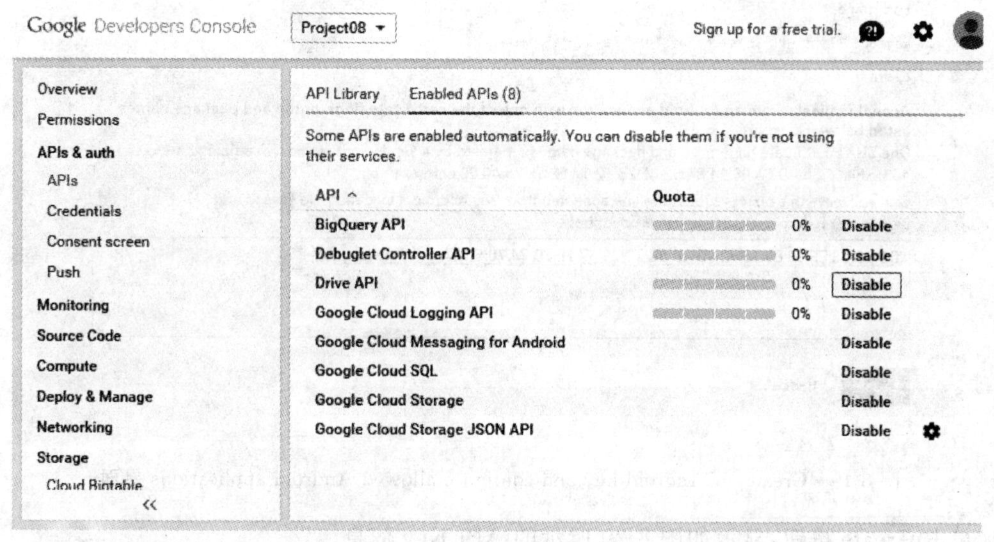

图 8-17 激活的 API 列表

(5) 单击左侧的 Credentials，这时分以下两种情况。

① 应用需要提交授权

方法是：在 OAuth 中单击 Create new Client ID 按钮，在弹出的对话框中选择 Installed application，在 Installed application type 中选择 Android，在 Package name 输入框中输入应用项目的包名，在 Signing certificate fingerprint (SHA1) 输入框中输入上一步得到的 SHA1，单击 Create Client ID 按钮，显示如图 8-18 所示的 API Key 信息。

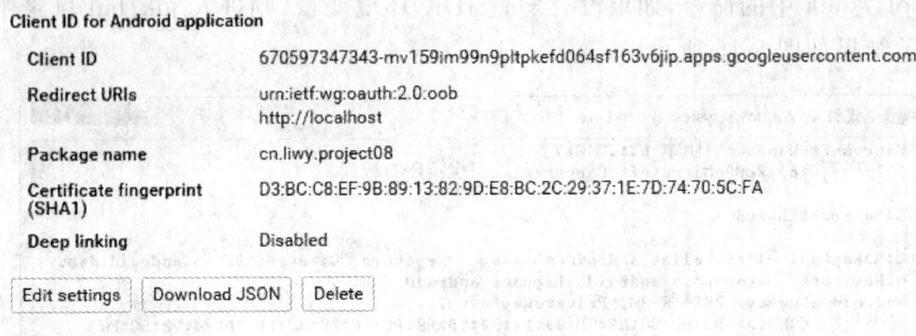

图 8-18　授权 API Key 获取界面

② 应用只是调用 API,不需要授权

在 Public API access 中单击 Create new Key 按钮,在弹出的对话框中选择 Android key,在文本输入框中输入上一步得到的 SHA1 并以英文分号结束,再加上应用程序包的名称,如图 8-19 所示。

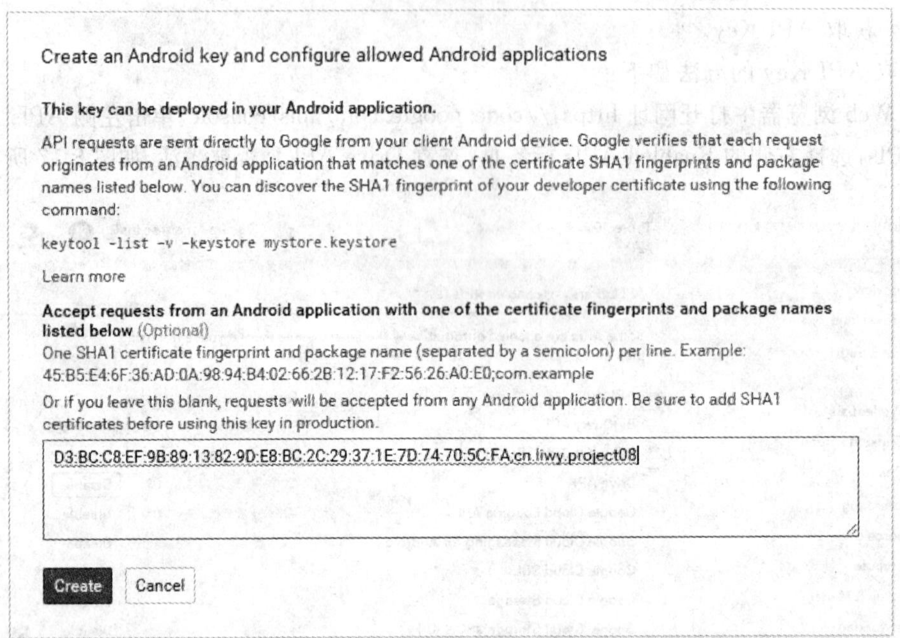

图 8-19　Create an Android key and configure allowed Android applications 界面

单击 Create 按钮,显示如图 8-20 所示的 API Key 信息。

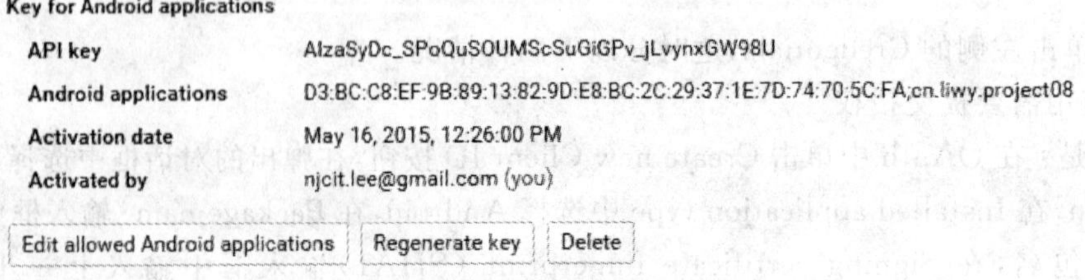

图 8-20　非授权 API Key 获取界面

8.3.2 创建授权 Google Drive 应用

下面基于 Google Drive for Android 创建一个应用，主界面授权登录成功后，以 ListView 的形式显示网络磁盘的文件。设计效果如图 8-21 所示。

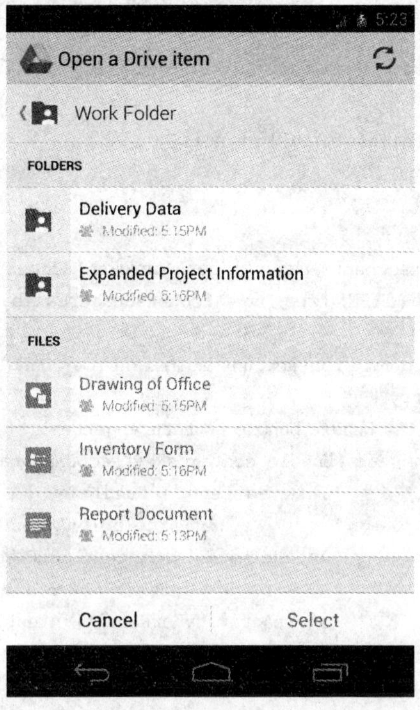

图 8-21 Google Drive 应用

步骤如下：

(1) 创建 Android 项目，并在 AndroidManifest.xml 中添加如下权限信息：

```
<uses-permission android:name="android.permission.INTERNET" />
<uses-permission android:name="android.permission.WRITE_EXTERNAL_STORAGE" />
<uses-permission android:name="android.permission.GET_ACCOUNTS" />
<uses-permission android:name="android.permission.USE_CREDENTIALS" />
<uses-permission android:name="android.permission.MANAGE_ACCOUNTS" />
<uses-permission android:name="android.permission.READ_GSERVICES" />
```

(2) 在 Activity 的 onCreate()生命周期方法中实现授权认证，代码如下：

```
settings = getPreferences(MODE_PRIVATE);
accountName = settings.getString(PREF_ACCOUNT_NAME, null);
credential.setAccessToken(settings.getString(PREF_AUTH_TOKEN, null));
Logger.getLogger("com.google.api.client").setLevel(LOGGING_LEVEL);
accountManager = new GoogleAccountManager(this);
gotAccount();
```

首先从 SharedPreferences 中获取以往的授权信息，初始化 GoogleAccountManager，然后调用 gotAccount()方法进行授权认证，代码如下：

```
void gotAccount() {
    Account account = accountManager.getAccountByName(accountName);
    if (account == null) {
        chooseAccount();
        return;
    }
    if (credential.getAccessToken() != null) {
        onAuthToken();
        return;
    }
    accountManager.getAccountManager().getAuthToken(account,
            AUTH_TOKEN_TYPE, true, new AccountManagerCallback<Bundle>() {

                public void run(AccountManagerFuture<Bundle> future) {
                    try {
                        Bundle bundle = future.getResult();
                        if (bundle.containsKey(AccountManager.KEY_INTENT)) {
                            Intent intent = bundle
                                    .getParcelable(AccountManager.KEY_INTENT);
                            intent.setFlags(intent.getFlags()
                                    & ~Intent.FLAG_ACTIVITY_NEW_TASK);
                            startActivityForResult(intent,
                                    REQUEST_AUTHENTICATE);
                        } else if (bundle
                                .containsKey(AccountManager.KEY_AUTHTOKEN)) {
                            setAuthToken(bundle
                                    .getString(AccountManager.KEY_AUTHTOKEN));
                            onAuthToken();
                        }
                    } catch (Exception e) {
                        Log.e(TAG, e.getMessage(), e);
                    }
                }
            }, null);
}
```

如果授权成功，调用 onAuthToken()方法异步加载 Google Drive 中的文件。

(3) 实现异步加载文件的 AsyncTask 实例，核心代码如下：

```
public class AsyncLoadFiles extends AsyncTask<Void, Void, List<String>> {

    ...
    public AsyncLoadFiles(GoogleDriveActivity driveSample) {
        this.driveSample = driveSample;
        service = driveSample.service;
        dialog = new ProgressDialog(driveSample);
```

```
    }

    @Override
    protected void onPreExecute() {
        dialog.setMessage("Loading root...");
        dialog.show();
    }

    @Override
    protected List<String> doInBackground(Void... arg0) {
        try {
            About about = service.about().get().execute();

            List<String> result = new ArrayList<String>();
            rootFile = service.files().get(about.getRootFolderId()).execute();
            result.add(rootFile.getTitle());
            return result;
        } catch (IOException e) {
            driveSample.handleGoogleException(e);
            return Collections.singletonList(e.getMessage());
        } finally {
            driveSample.onRequestCompleted();
        }
    }

    @Override
    protected void onPostExecute(List<String> result) {
        dialog.dismiss();
        driveSample.setListAdapter(new ArrayAdapter<String>(driveSample,
                android.R.layout.simple_list_item_1, result));
        driveSample.getListView().setOnItemClickListener(
                new OnItemClickListener() {

                    @Override
                    public void onItemClick(AdapterView<?> arg0, View arg1,
                            int arg2, long arg3) {
                        Intent intent = new Intent(driveSample,
                                DocListActivity.class);
                        intent.putExtra(DocListActivity.KEY_ROOT_FOLDER_ID,
                                rootFile.getId());
                        intent.putExtra(DocListActivity.KEY_AUTH_TOKEN,
                                driveSample.credential.getAccessToken());
                        driveSample.startActivity(intent);
                    }
                });
    }
}
```

其中,19～24 行获取 Google Drive 根路径下的文件集合,然后通过 33～53 行的 onPostExecute()回调方法,将集合数据提交给基于 BaseAdapter 的 DocListAdapter 适配器,最终将文件信息显示在 ListView 中。同时,该方法实现了 ListView 中每个 Item 的单击事件,即跳转到 DocListActivity 界面,如果单击的是文件夹,则加载该 Item 的子文件夹信息。

如果单击的是文件,则图 8-22 演示了 Drive File 操作的生命周期流程,有关 Google Drive API 的详细信息,请参阅 https://developer.android.com/reference/com/google/android/gms/drive/package-summary.html。

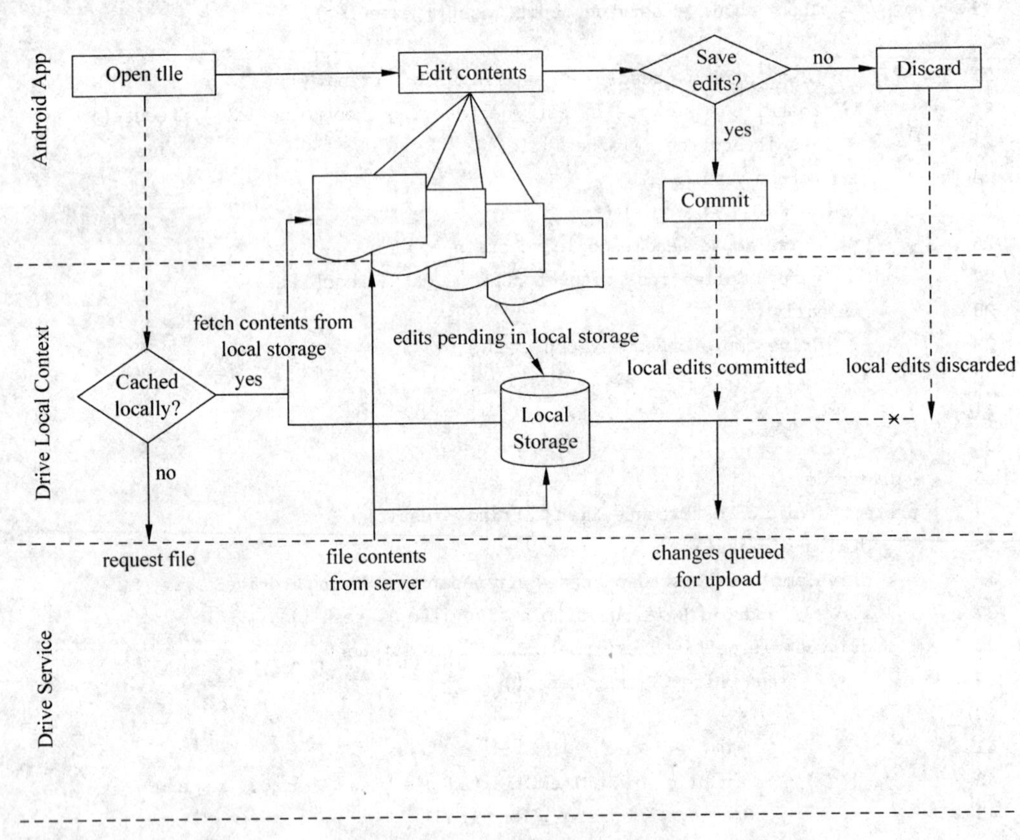

图 8-22　Drive File 操作的生命周期

8.4　习　　题

1. 编程实现基于 BackupAgentHelper 的文件备份功能。
2. 编程实现基于 Google Drive 的文本文档的在线编辑功能。

第 9 章　Philm 项目分析与设计

Philm[①] 是 Chris Banes 在 GitHub 上发布的一个关于电影信息的开源 Android 应用。项目基于 Apache v2.0 许可协议，使用 Trakt[②] 和 TMDB[③] 开放接口。

9.1　应用简介

启动应用后，显示如图 9-1 所示的主界面，主要是侧边栏中“发现”模块的信息。

- 流行：显示最近观看和关注较多的影片信息。
- 上映中：正在热映的影片信息。
- 即将上映：即将上映的影片信息。
- 推荐：Trakt 推荐的影片信息。

以上四个板块实现拖动刷新。单击界面左上角的☰按钮，激活侧边栏，包括登录、发现、我的收藏夹、观看列表和搜索菜单，如图 9-2 所示。

图 9-1　应用主界面

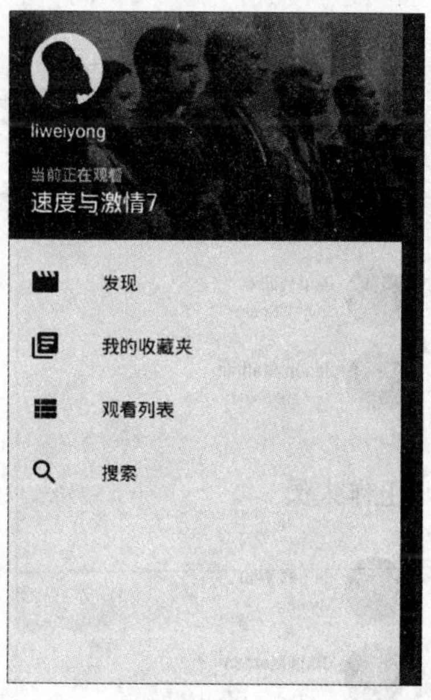

图 9-2　侧边菜单

---

① 项目主页为 https://github.com/chrisbanes/philm。

② Trakt(https://trakt.tv/)是一个在线电视剧、电影媒体资源追踪平台，用户可以通过多设备工具来使用 Trakt 平台发现自己喜欢的电影，追踪最新的内容动态，预定电视剧最新剧集。

③ TMDB(https://www.themoviedb.org/)是一个提供电影、演员等信息的在线数据库。

单击主界面的任一电影海报,启动电影详情界面,如图 9-3～图 9-6 所示。详情界面包括电影简介、电影详情、观影留言、添加到收藏夹、添加到观影列表、评分等。另外,还包括电影预告片、演员及工作人员信息等。

图 9-3　电影详情界面 1

图 9-4　电影详情界面 2

图 9-5　电影详情界面 3

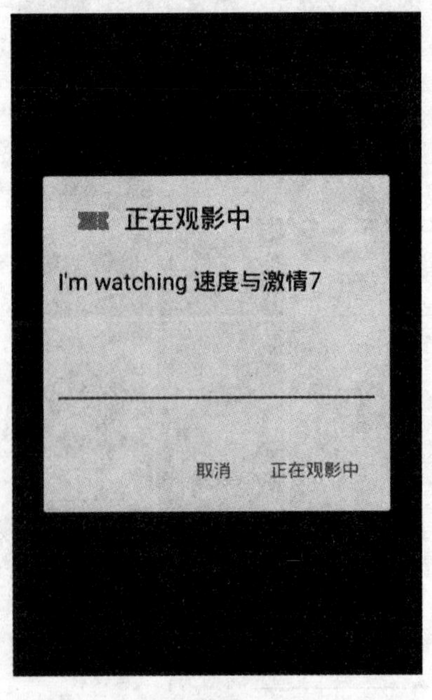

图 9-6　观影留言界面

单击电影详情界面的演员列表,显示演员详细信息界面,如图 9-7 所示。另外,可以看到和演员相关的工作人员信息。

单击侧边栏中的登录选项，显示如图 9-8 所示的登录/注册界面。登录成功后，可以在侧边栏的“我的收藏夹”选项中查看收藏的电影信息，包括已看过和未看过的电影信息等，如图 9-9 所示。

图 9-7　演员详情界面

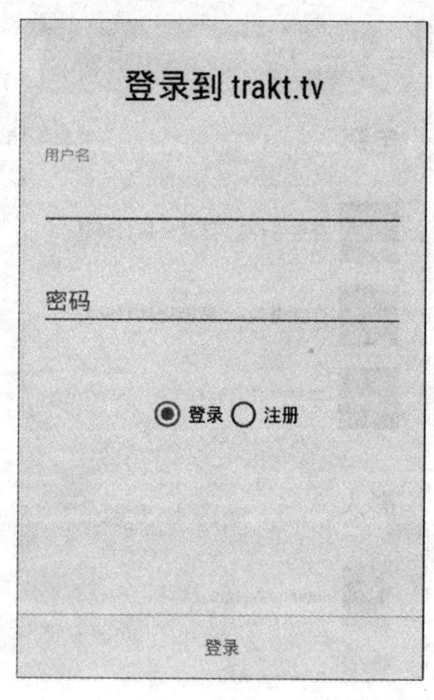

图 9-8　登录界面

单击侧边栏中的观看列表选项，显示如图 9-10 所示的界面，其中包含了当前用户的观影列表记录。单击其中的影片，显示如图 9-1 所示的电影详情界面。

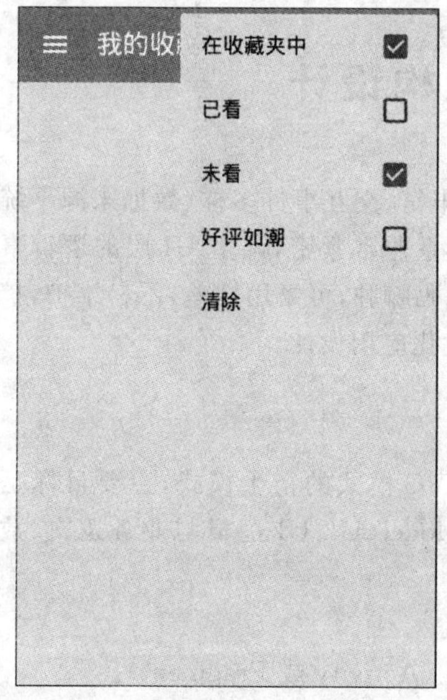

图 9-9　我的收藏界面

图 9-10　观看列表界面

单击侧边栏中的搜索选项，显示如图 9-11 所示的搜索界面，在上方的搜索关键词输入框中输入搜索信息后，在下方显示相关电影和演员信息，还可以显示如图 9-12 所示的相关电影信息。

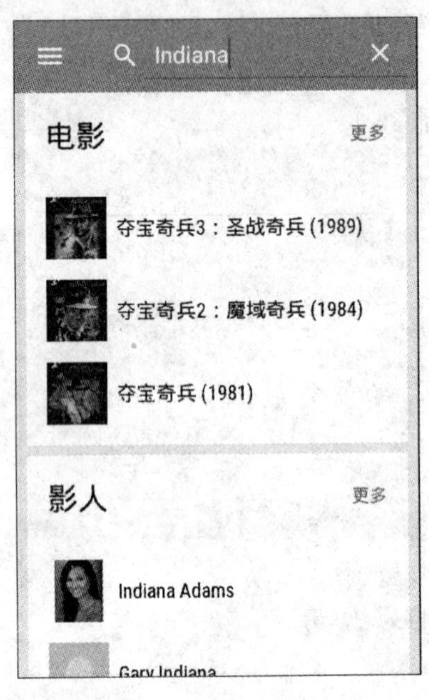

图 9-11 搜索界面

图 9-12 相关电影列表

除了上面介绍的功能外，应用还提供了批项目处理、界面刷新、系统设置以及应用信息等功能。

9.2 应用架构设计

通过 9.1 节展示的应用来看，Philm 具有 UI 丰富、交互事件多样、数据来源不统一等特点。传统的开发模式中，Activity 与 Model 层的关系非常紧密，许多 UI 层的逻辑事件都是由 Activity 来完成的，这样的设计使得 Activity 代码臃肿，可复用性差，不易维护，扩展性较差等。因此，需要合理的应用架构和设计模式来优化应用设计。

9.2.1 MVP 设计模式

MVP(Model-View-Presenter)设计模式是 MVC 模式的衍生模式，主要目标是将显示逻辑与业务逻辑分离，即不容许 View 直接访问 Model(UI、UI 逻辑与业务逻辑、数据三者隔离开来)，如图 9-13 所示。

在 Android 中，MVP 模式通常包含 4 个要素。

① View：负责绘制 UI 元素、与用户进行交互(Activity 或 Fragment)。

② View Interface：View 需要实现的接口，View 通过 View Interface 与 Presenter 进行交互，降低耦合，方便进行单元测试。

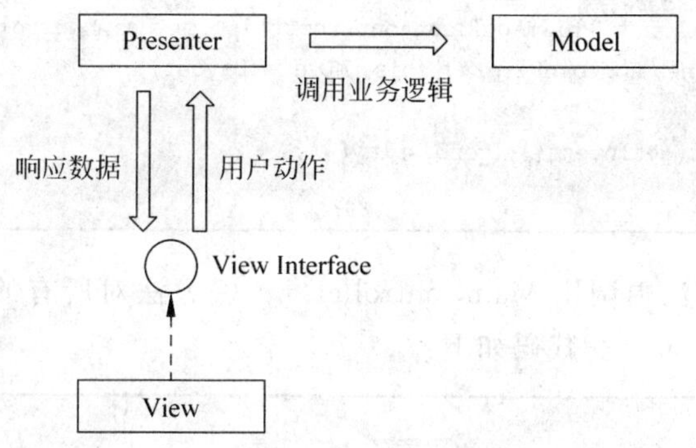

图 9-13　MVP 设计模式

③ Model：负责存储、检索、操纵数据(有时也通过一个 Model Interface 来降低耦合)。

④ Presenter：从 Model 中获取数据并提供给 View 层，Presenter 还负责处理后台任务，承载了大部分的复杂逻辑。

图 9-13 清晰地展示了 MVP 模式的几个特点：

- View 和 Model 完全解耦，两者不发生直接关联，通过 Presenter 进行通信，便于维护和测试。
- Presenter 并不是与具体的 View 耦合，而是和一个抽象的 View Interface 耦合，只需要在 Presenter 中为不同的 View 定义 View Interface 即可，具体的 View 实现自己的 View Interface，即可使用 Presenter 中的 Model 操作等。
- View 在 MVP 里应该是一个"极瘦"的概念，最多也只能包含维护自身状态的逻辑，而其他逻辑都应实现在 Presenter 中。

1. Philm 总体架构设计

Philm 使用 Controller 来统一管理 Model 和 View，该 Controller 相当于 MVP 中的 Presenter，在 Controller 内部，直接定义了 MVP 中 View 与 Presenter 的交互接口 Callback，另外该项目引入了 State 的概念，来统一管理 Model 和业务中所需的 Event，所有的 Activity 和 Fragment 的跳转以及 TitleBar 和 Drawer 的管理使用一个 Display 来实现。这些角色之间的关系如图 9-14 所示。

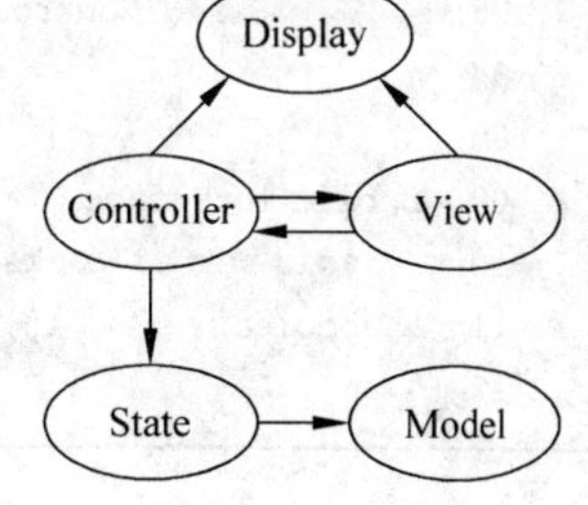

图 9-14　MVP 对象间的关系

下面分析一下 MainActivity 的实现流程。

(1) 初始化 Controller 和 Display。

在 Philm 中，View 由 Fragment 和 Activity 实现，由于应用启动时首先启动 Activity，因此，绑定 Presenter 的工作要在 View 中进行(也就是 Activity/Fragment)。MainActivity 继承自 BasePhilmActivity，在 BasePhilmActivity 的 onCreate()方法中初始化 MainController 和 Display。核心代码如下：

```
@Override
protected void onCreate(Bundle savedInstanceState) {
    ...
```

```
    mMainController = PhilmApplication.from(this).getMainController();
    mDisplay = new AndroidDisplay(this, mDrawerLayout);

    handleIntent(getIntent(), getDisplay());
}
```

在 onResume()方法中调用 MainController.init()方法对所有的 Controller 进行初始化、事件注册并设置 Display。代码如下：

```
@Override
protected void onResume() {
    super.onResume();
    mMainController.attachDisplay(mDisplay);
    mMainController.setHostCallbacks(this);
    mMainController.init();
}
```

最后在 onPause()中解除对上述资源的操作。

(2) 实现回调接口。

根据 MVP 的思想，Activity/Fragment 是一个 View(UI)，UI 的逻辑由 Controller 通过持有 UI 的 UC 来处理(UC 用来响应 UI 的动作)。MainActivity 对应的 UI 为 MainController.MainUi，代码如下：

```
public interface MainUi extends MainControllerUi {
    void showLoginPrompt();
}

public interface MainControllerUi extends
        BaseUiController.Ui<MainControllerUiCallbacks> {
}

public interface Ui<UC> {
    void setCallbacks(UC callbacks);
    boolean isModal();
}
```

MainUi 总共有三个方法：

- void showLoginPrompt()
- void setCallbacks(UC callbacks)
- boolean isModal()

MainUi 对应的 UC 为 MainControllerUiCallbacks。

(3) 渲染 UI。

在相应的 Fragment 的 onResume()方法中调用 attachUi()方法为 Fragment 添加所有的 Callback 并进行视图的渲染。BaseMovieControllerListFragment 的 onResume()方法代码如下：

```java
@Override
public void onResume() {
    super.onResume();
    getController().attachUi(this);
}
```

attachUi()的实现在 BaseUiController 中，代码如下：

```java
public synchronized final void attachUi(U ui) {
    Preconditions.checkArgument(ui != null, "ui cannot be null");
    Preconditions.checkState(!mUis.contains(ui), "UI is already attached");

    mUis.add(ui);

    ui.setCallbacks(createUiCallbacks(ui));

    if (isInited()) {
        if (!ui.isModal() && !(ui instanceof SubUi)) {
            updateDisplayTitle(getUiTitle(ui));
        }

        onUiAttached(ui);
        populateUi(ui);
    }
}
```

可以看出，attachUi()方法通过 ui. setCallbacks(createUiCallbacks(ui))设置回调，通过 populateUi(ui)绘制 UI。在 MainController 中，createUiCallbacks(ui)的实现代码如下：

```java
@Override
protected MainControllerUiCallbacks createUiCallbacks(
        final MainControllerUi ui) {
    return new MainControllerUiCallbacks() {
        @Override
        public void onSideMenuItemSelected(SideMenuItem item) {
            Display display = getDisplay();
            if (display != null) {
                showUiItem(display, item);
                display.closeDrawerLayout();
            }
        }

        @Override
        public void addAccountRequested() {
            Display display = getDisplay();
            if (display != null) {
                display.startAddAccountActivity();
                display.closeDrawerLayout();
```

```
            }
        }

        @Override
        public void showMovieCheckin() {
            Display display = getDisplay();
            WatchingMovie checkin = mState.getWatchingMovie();

            if (display != null && checkin != null) {
                display.closeDrawerLayout();
                display.startMovieDetailActivity(checkin.movie.getImdbId(),
                        null);
            }
        }

        @Override
        public void setShownLoginPrompt() {
            mPreferences.setShownTraktLoginPrompt();
        }
    };
}
```

Philm 设置 View 回调采用了一个非常精巧的设计：通过泛型来处理 ViewInterface 与回调。BaseUiController 的泛型可以表示为 BaseUiController <ViewInterface, ViewCallbacks>，其中前一个泛型代表了用于解耦 View 和 Presenter 而实现的 ViewInterface，而后一个是提供给 ViewInterface 的回调。ViewInterface 本身提供了 setCallbacks 方法来进行 ViewCallbacks 与 View 的绑定。这样，所有的业务逻辑都可以在 Presenter 中实现，之后通过回调提供给 View。View 仅仅负责 UI 的表现，在需要处理业务逻辑时，通过调用之前 Presenter 设置的回调实现。例如，单击网格中的影片海报时，一般回调处理如下：

```
@Override
public void onListItemClick(GridView l, View v, int position, long id) {
    if (hasCallbacks()) {
        ListItem<PhilmMovie> item = (ListItem<PhilmMovie>) l
                .getItemAtPosition(position);
        if (item.getListType() == ListItem.TYPE_ITEM) {
            getCallbacks().showMovieDetail(item.getListItem(),
                    ActivityTransitions.scaleUpAnimation(v));
        }
    }
}
```

在渲染视图时，从 State 中获取 Bean，如果没有启动子线程获取，则发送一个 Event 到 Controller 进行视图更新。例如 MainController 的 populateUi()方法代码如下：

```
private void populateUi(SideMenuUi ui) {
    ui.setSideMenuItems(getEnabledSideMenuItems(),
            mState.getSelectedSideMenuItem());

    PhilmUserProfile profile = mState.getUserProfile();
    if (profile != null) {
        ui.showUserProfile(profile);
    } else {
        ui.showAddAccountButton();
    }

    WatchingMovie checkin = mState.getWatchingMovie();
    if (checkin != null) {
        ui.showMovieCheckin(checkin);
    } else {
        ui.hideMovieCheckin();
    }
}
```

(4) 实现 View。

MainActivity 的布局是一个 DrawerLayout，包含 MenuFragment(SideMenuFragment)和显示内容的 Fragment。首次展示的 Fragment 是通过 BasePhilmActivity 中 handleIntent(Intent intent, Display display)来确定的，代码如下：

```
@Override
protected void handleIntent(Intent intent, Display display) {
    if (Intent.ACTION_MAIN.equals(intent.getAction())) {
        if (!display.hasMainFragment()) {
            getMainController().setSelectedSideMenuItem(
                    MainController.SideMenuItem.DISCOVER);
            display.showDiscover();
        }
    }
}
```

在 Philm 项目中，通过查看 AndroidDisplay 可以知道，整个项目的 Activity 跳转、Fragment 切换、DrawerLayout 的开和闭、ActionBar 的控制都是通过 Display 来实现的。通过查看 display. showDiscover()可以发现，Mainactivity 中的 Fragment 对应的类为 DiscoverTabFragment，其主要功能是使用一个 ViewPager 管理几个 Fragment。

以上 MVP 组件之间的相互调用如图 9-15 所示。

2. Presenter 与 Display 设计

Presenter 是控制中心，统一管理界面状态的初始化和状态清理，为所有的 UI 进行渲染并添加 CallBack，统一调度业务相关的后台任务线程，订阅 State 中定义的 Event，进行视图的更新。Presenter 统一定义了 View 与 Presenter 的 View Interface，一个 Presenter 可以为多个 View 定义 View Interface，因此 Presenter 可以被多个 View 所共用。

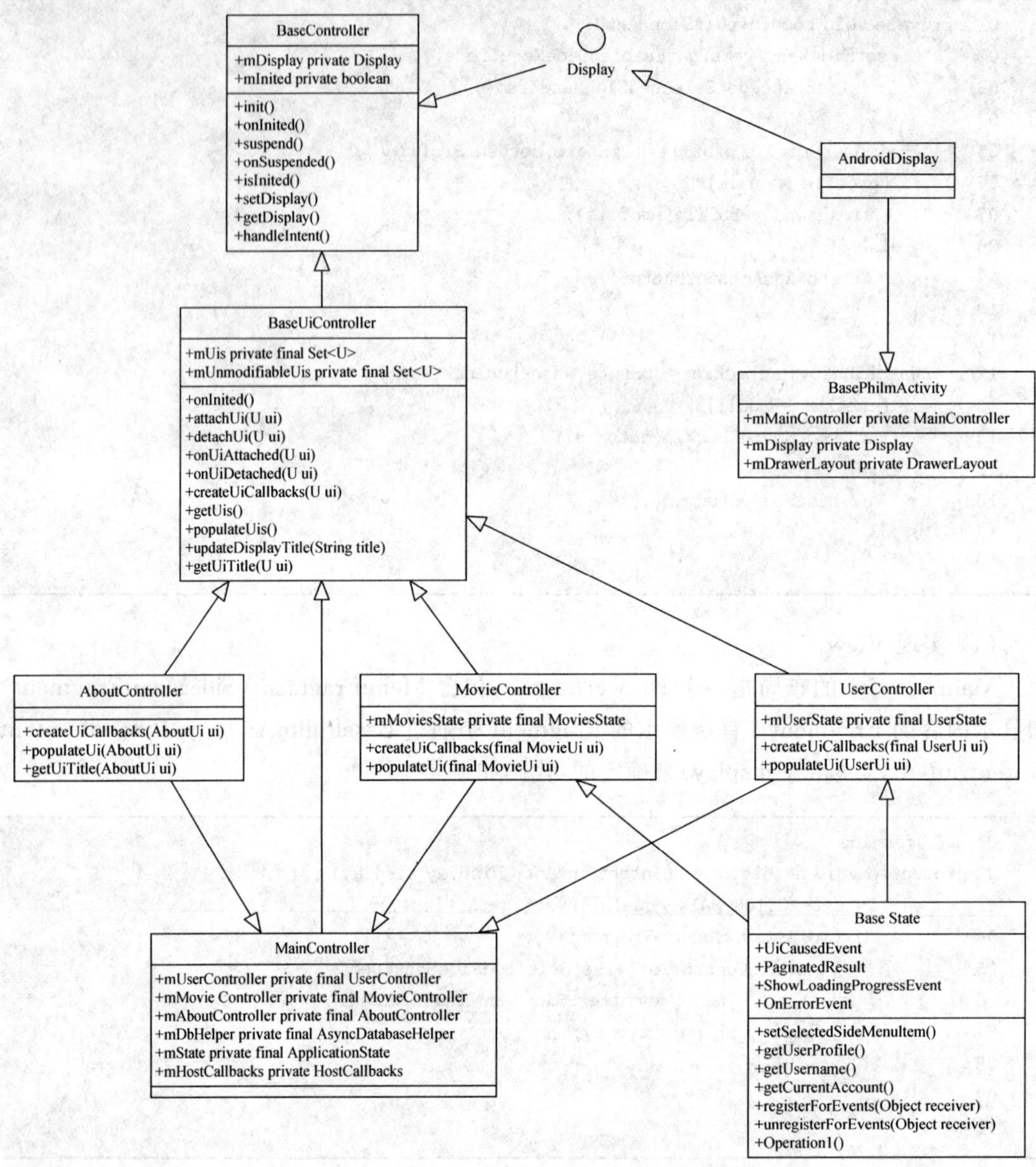

图 9-15　MVP 的类关系图

Philm 的 Presenter 被设计成多个 Controller,包括 AboutController、BaseController、BaseUiController、MainController、MovieController 和 UserController。

BaseController 核心代码如下:

```
abstract class BaseController {
    ...

    public final void init() {
        Preconditions.checkState(mInited == false, "Already inited");
        mInited = true;
```

```
        onInited();
    }

    public final void suspend() {
        Preconditions.checkState(mInited == true, "Not inited");
        onSuspended();
        mInited = false;
    }
}
```

init()是所有 Controller 初始化发起的方法，例如，在 BasePhilmActivity 中通过 mMainController.init()发起，所有的 Controller 由 MainCtroller 统一控制。suspend()是销毁 Controller 的地方。

BaseUiController 统一定义管理某个界面的所有的 UI、事件和接口，核心代码如下：

```
abstract class BaseUiController<U extends BaseUiController.Ui<UC>, UC>
        extends BaseController {
    ...
    public interface Ui<UC> {
        void setCallbacks(UC callbacks);
        boolean isModal();
    }

    @Inject Logger mLogger;
    private final Set<U> mUis;
    private final Set<U> mUnmodifiableUis;

    public BaseUiController() {
        mUis = new CopyOnWriteArraySet<>();
        mUnmodifiableUis = Collections.unmodifiableSet(mUis);
    }

    public synchronized final void attachUi(U ui) {
        Preconditions.checkArgument(ui != null, "ui cannot be null");
        Preconditions.checkState(!mUis.contains(ui),
                "UI is already attached");
        mUis.add(ui);
        ui.setCallbacks(createUiCallbacks(ui));

        if (isInited()) {
            if (!ui.isModal() && !(ui instanceof SubUi)) {
                updateDisplayTitle(getUiTitle(ui));
            }

            onUiAttached(ui);
            populateUi(ui);
        }
    }
```

```

    protected final void updateDisplayTitle(String title) {
        Display display = getDisplay();
        if (display != null) {
            display.setActionBarTitle(title);
        }
    }

    public synchronized final void detachUi(U ui) {
        Preconditions.checkArgument(ui != null, "ui cannot be null");
        Preconditions.checkState(mUis.contains(ui), "ui is not attached");
        onUiDetached(ui);
        ui.setCallbacks(null);

        mUis.remove(ui);
    }

    protected void onInited() {
        if (!mUis.isEmpty()) {
            for (U ui : mUis) {
                onUiAttached(ui);
                populateUi(ui);
            }
        }
    }

    protected synchronized final void populateUis() {
        if (Constants.DEBUG) {
            mLogger.d(getClass().getSimpleName(), "populateUis");
        }
        for (U ui : mUis) {
            populateUi(ui);
        }
    }

    protected synchronized U findUi(final int id) {
        for (U ui : mUis) {
            if (getId(ui) == id) {
                return ui;
            }
        }
        return null;
    }

    protected final void populateUiFromEvent(
                BaseState.UiCausedEvent event) {
        Preconditions.checkNotNull(event, "event cannot be null");

        final U ui = findUi(event.callingId);
```

```
83          if (ui != null) {
84              populateUi(ui);
85          }
86      }
87
88  }
```

每个 Controller 都拥有一个自己的 UiCallBacks，以及一个继承自 BaseUiController. Ui 的 UI。

在 Controller 内部也可同时为不同的 View 定义相应的 UICallBack，因为不同的 View 可能会使用到相同的 Model、State 等事件，不同的 View 只需实现 Controller 中定义的与自己相关的 CallBack 即可。当然该 UICallBack 需要继承已实现 BaseUiController. Ui 的 UI，因为最终都会传入 BaseUIController 中进行 setCallbacks(UC)。

04～07 行的 UI＜UC＞接口是所有 Controller 的子类所在 UI 都需要实现的 UI 接口，拥有 CallBack 的能力，在 Controller 调用 attachUi()时会为传入的 UI 设置 CallBack。

10 行的 mUis 用于存储所有的 UI。11 行的 mUnmodifiableUis 是 mUis 的复制，但不可修改。

18～33 行的 attachUi(U ui)方法在 Activity 或者 Fragment 的 onResume()中调用，进行添加回调、视图渲染等工作。该方法是 final 类型，不可复写，只是为了 Controller 的实现类进行状态的管理，如果需要更多的操作，可以实现 onUiAttached(UI)方法，该方法会在 attachUI 中调用。

51～58 行的 onInited()方法在各个 Controller 进行初始化时调用，不用明确地调用，因为 MainController 统一管理了所有的 Controller，例如，在 BasePhilmActivity 的 onResume()中通过调用 MainController. init()方法统一初始化。

78～86 行的 populateUiFromEvent(BaseState. UiCausedEvent event)方法在某些数据更新，需要更新视图时调用该方法，发送一个事件，通知 Controller 进行 UI 更新。

Display 是 Philm 中一处比较新颖的设计，它本身应该算是 Presenter 的一部分，具体的实现在 AndroidDisplay 类中。AndroidDisplay 的作用是负责应用中所有 UI 切换的工作，比如 Activity 跳转、现实 Fragment、弹出 Drawer 等。AndroidDisplay 内部没有业务逻辑操作。AndroidDisplay 核心代码如下：

```
public class AndroidDisplay implements Display {
    ...
    public AndroidDisplay(ActionBarActivity activity,
            DrawerLayout drawerLayout) {
        mActivity = Preconditions.checkNotNull(activity,
                "activity cannot be null");
        mDrawerLayout = drawerLayout;

        mActivity.getTheme().resolveAttribute(R.attr.colorPrimaryDark,
                sTypedValue, true);
        mColorPrimaryDark = sTypedValue.data;

```

```
        if (mDrawerLayout != null) {
            mDrawerLayout.setStatusBarBackgroundColor(mColorPrimaryDark);
        }
    }

    @Override
    public void showLibrary() {
        showFragmentFromDrawer(new LibraryMoviesFragment());
    }

    @Override
    public void startMovieDetailActivity(String movieId, Bundle bundle) {
        Intent intent = new Intent(mActivity, MovieActivity.class);
        intent.putExtra(PARAM_ID, movieId);
        startActivity(intent, bundle);
    }


    @Override
    public void closeDrawerLayout() {
        if (mDrawerLayout != null &&
              mDrawerLayout.isDrawerOpen(GravityCompat.START)) {
            mDrawerLayout.closeDrawers();
        }
    }
}
```

在 BasePhilmActivity 中调用 mDisplay = new AndroidDisplay(this, mDrawerLayout)初始化 Display,然后在 MainController 中调用 showUiItem()等方法控制 UI 的显示。

3. Model 与 State 设计

Philm 中的 Model 通过 State 来管理,State 保存界面使用到了业务 Bean,定义业务中用到了 Event 事件,包括异步请求完成的通知、数据变更、NetWork 的状态变化、LoadingProgress 的展示与隐藏等。注意,State 并不对 Model 做复杂的操作,只是简单的 Set 和 Get 操作,复杂的操作全部由 Controller 处理。

每一个界面都拥有自己的 State 接口,ApplicationState 负责统一实现所有 State 接口,是一个包含了所有 Model 总和的类,扮演了 UI 和后台线程的通信者,实现保存业务 Bean 的分发和事件的分发。ApplicationState 核心代码如下:

```
public final class ApplicationState implements BaseState,
        MoviesState, UserState {
    ...

    public ApplicationState(Bus eventBus) {
        mEventBus = Preconditions.checkNotNull(eventBus,
                "eventBus cannot null");
        mTmdbIdMovies = new ArrayMap<>(INITIAL_MOVIE_MAP_CAPACITY);
```

```
        mImdbIdMovies = new ArrayMap<>(INITIAL_MOVIE_MAP_CAPACITY);
        mPeople = new ArrayMap<>();
    }

    @Override
    public void registerForEvents(Object receiver) {
        mEventBus.register(receiver);
    }

    @Override
    public Map<String, PhilmMovie> getTmdbIdMovies() {
        return mTmdbIdMovies;
    }

    @Override
    public void setLibrary(List<PhilmMovie> items) {
        if (!Objects.equal(items, mLibrary)) {
            mLibrary = items;
            mEventBus.post(new LibraryChangedEvent());
        }
    }
}
```

在使用 Controller 进行界面渲染时，首先从 State 中获取业务 Bean。如果 State 中没有，Controller 则会启动后台线程请求数据，成功获取数据后通过 state.set()分发保存到 State 中，同时会发出一个 Event 通知 Controller 进行相应的处理。

9.2.2 Dagger 与依赖注入

1. 依赖注入

依赖注入(Dependency Injection)又称控制反转(Inversion of Control)，是一种软件设计模式，其基本思想是：用一个单独的对象获得接口的一个合适的实现，并将其实例赋给调用者的一个字段。当某个角色(可能是一个 Java 实例，调用者)需要另一个角色(另一个 Java 实例，被调用者)的协助时，在传统的程序设计过程中，通常由调用者来创建被调用者的实例。如果创建被调用者实例的工作不再由调用者来完成，而是由外部容器完成，则称为控制反转。创建被调用者实例的工作通常由外部容器来完成，然后注入调用者，因此也称为依赖注入。

依赖注入涉及多个元素：

- 调用者(依赖者)；
- 一份依赖的声明，以接口方式定义；
- 一个注入器(有时也叫提供者、容器)，它能创建实现了接口(定义依赖的接口)的类的实例。

调用者会描述它需要哪些被调用者才能正常工作。再由注入容器来决定哪些具体的实现能满足调用者的需求，并提供给调用者。

在传统的软件开发中，调用者需要自己来确定被调用的对象。但是在依赖注入模式中，这个决定权授权给了注入容器,注入容器能在软件运行时选择替换不同的实现,而不是在编译时。这也是依赖注入的关键优势。

依赖注入有三种典型的方式。

(1) 构造器注入(Constructor Injection):注入容器会智能地选择和调用适合的构造方法以创建依赖的对象。如果被选择的构造方法具有相应的参数,注入容器在调用构造方法之前解析注册的依赖关系并自行获得相应参数对象。

(2) 属性注入(Property Injection):如果需要使用到被依赖对象的某个属性,在创建被依赖对象之后,注入容器会自动初始化该属性。

(3) 方法注入(Method Injection):如果被依赖对象需要调用某个方法进行相应的初始化,在该对象创建之后,注入容器会自动调用该方法。

2. Dagger[①] 框架

Dagger 是由专注于移动支付的公司——Square 公司推出的一种 Android 平台的依赖注入框架。Dagger 构建在标准的 javax. inject Annotation 基础之上,使用 Annotation 给需要注入的对象做标记,通过 inject()方法自动注入所有对象,从而完成自动的初始化。Dagger 的框架如图 9-16 所示。

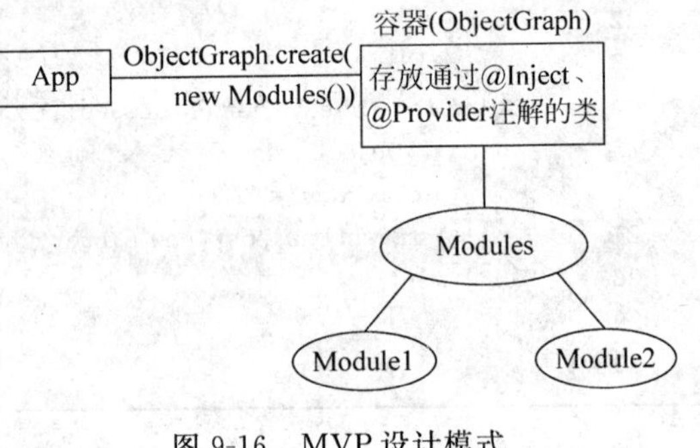

图 9-16 MVP 设计模式

Dagger 使用的基本步骤如下。

(1) 在相关类的构造方法前添加一个@Inject 注解,Dagger 就会在需要获取该对象时,调用这个被标记的构造方法,从而生成一个对象实例。例如:

```
public class MusicBean {
    ...

    @Inject
    public MusicBean() {
        ...
    }

    ...
}
```

注意:如果构造方法含有参数,Dagger 会在调用构造对象时先去获取这些参数,所以要保证它的参数也提供可被 Dagger 调用到的生成方法。Dagger 可调用的对象生成方式有两种:一种是用@Inject 修饰的构造方法,上面就是这种方式。另外一种是用@Provides 修饰的方法。

(2) 在 Activity 中注入类实例。具体方法是在 Activity 中的 MusicBean 属性声明之前

① Dagger 官网:http://square. github. io/dagger/。

添加@Inject 注解，这样 Dagger 就知道哪些属性需要被注入，然后调用通过@Inject 注解的构造方法。例如：

```
public class MainActivity extends Activity {
    @Inject MusicBean musicBean;
    ...
}
```

对构造方法进行注解是很好的实现依赖的途径，然而它并不适用于所有情况。

- 接口没有构造方法，不能被构造。
- 第三方的类不能被注释构造。
- 有些类需要灵活选择初始化的配置，而不是使用一个单一的构造方法。

(3) 在 Activity 的合适位置（一般在 onCreate( ) 生命周期回调方法中）调用 ObjectGraph. inject()方法，Dagger 就会自动调用步骤(1)中的生成方法生成依赖的实例，并注入当前对象中（如 MainActivity）。

```
public class MainActivity extends Activity {
    @Inject MusicBean musicBean;

    @Override
    protected void onCreate(Bundle savedInstanceState) {
        ObjectGraph.create(AppModule.class).inject(this);
    }
    ...
}
```

这样，Android 开发工具 apt 会在 MainActivity 所在的 package 下生成一个辅助类 MainActivity $ $ InjectAdapter，并在该类中实现 injectMembers()方法，代码类似如下：

```
public void injectMembers(MainActivity paramMainActivity) {
    paramMainActivity.musicBean = ((MusicBean)musicBean.get());
    ...
}
```

上面通过 ObjectGraph. inject()方法传入 paramMainActivity，并且 musicBean 属性是 package 权限，所以 Dagger 只需要调用这个辅助类的 injectMembers()方法即可完成依赖注入，这里的 musicBean. get()会调用 MusicBean 的生成方法。

至此，就完成了使用 Dagger 的@Inject 方式将一个 MusicBean 对象注入 MainActivity 的流程。

3. Philm 中的依赖注入

在 Philm 项目中，复杂的用户界面和事件逻辑需要多处获取账户信息、网络状态等，如果在每次使用这些对象时都通过 new 方法来获取，将使代码的耦合度非常高，给后期的维护带来很大的麻烦。下面结合 Philm 项目，具体分析使用 Dagger 实现依赖注入模式来解耦合代码，各层对象的调用完全面向接口，这样，当系统重构时，代码的改写量将大大减少。

下面是账户管理的依赖注入过程。

(1) 定义账户管理接口,代码如下:

```
public interface PhilmAccountManager {
    public List<PhilmAccount> getAccounts();
    public void addAccount(PhilmAccount account);
    public void removeAccount(PhilmAccount account);
    public void updatePassword(PhilmAccount account);
}
```

(2) 实现账户管理接口,核心代码如下:

```
public class AndroidAccountManager implements PhilmAccountManager {

    private final AccountManager mAccountManager;

    public AndroidAccountManager(AccountManager accountManager) {
        mAccountManager = Preconditions.checkNotNull(accountManager,
                "accountManager cannot be null");
    }

    @Override
    public List<PhilmAccount> getAccounts() {
        final Account[] accounts = mAccountManager
                .getAccountsByType(Constants.TRAKT_ACCOUNT_TYPE);
        ArrayList<PhilmAccount> philmAccounts = new
                ArrayList<>(accounts.length);

        for (int i = 0; i < accounts.length ; i++) {
            final Account account = accounts[i];

            String password = mAccountManager.getPassword(account);
            philmAccounts.add(new PhilmAccount(account.name, password));
        }

        return philmAccounts;
    }

    @Override
    public void addAccount(PhilmAccount philmAccount) {
        ...
    }

    @Override
    public void removeAccount(PhilmAccount philmAccount) {
        ...
    }

```

```
    @Override
    public void updatePassword(PhilmAccount philmAccount) {
        ...
    }

}
```

(3) 实现依赖关系，代码如下：

```
@Module(
        includes = ContextProvider.class,
        library = true
)
public class AccountsProvider {

    @Provides @Singleton
    public PhilmAccountManager provideAccountManager(AccountManager
            androidAccountManager) {
        return new AndroidAccountManager(androidAccountManager);
    }

}
```

说明：

- 一个 Module 中所有@Provides 方法的参数都必须在这个 Module 中提供相应的@Provides 方法，或者在@Module 注解后添加“complete = false”，注明这是一个不完整 Module(即它会被其他 Module 所扩展)。
- 一个 Module 中所有的@Provides 方法都要被它声明的注入对象所使用，或者在@Module 注解后添加“library = ture”注明(即它是为了扩展其他 Module 而存在的)。
- 通常情况下，约定@Provides 方法以 provide 作为前缀，@Module 类以 Module 作为后缀。
- 所有带有@Provides 注解的方法都需要被封装到带有@Module 注解的类中。
- @Singleton 是一种单例注解，通过添加@Singleton 注解之后，对象只会被初始化一次，之后的每次都会被直接注入相同的对象。

(4) 构建全局依赖关系，代码如下：

```
@Module(
        injects = PhilmApplication.class,
        includes = {
                UtilProvider.class,
                AccountsProvider.class,
                NetworkProvider.class,
                StateProvider.class,
                PersistenceProvider.class,
```

```
                InjectorModule.class
        }
)
public class ApplicationModule {
}
```

(5) 构建 ObjectGraph。ObjectGraph 是由 Dagger 提供的抽象工具类，负责 Dagger 所有的业务逻辑，Dagger 最关键流程都是从这个类发起的，包括依赖关系图创建、实例(依赖或宿主)获取、依赖注入。上一步的 PhilmApplication 代码如下：

```
public class PhilmApplication extends Application implements Injector {

    public static PhilmApplication from(Context context) {
        return (PhilmApplication) context.getApplicationContext();
    }

    @Inject MainController mMainController;

    private ObjectGraph mObjectGraph;

    @Override
    public void onCreate() {
        super.onCreate();

        if (AndroidConstants.STRICT_MODE) {
            StrictMode.setThreadPolicy(new StrictMode.ThreadPolicy.Builder()
                    .detectAll()
                    .penaltyLog()
                    .penaltyDialog()
                    .build());
            StrictMode.setVmPolicy(new StrictMode.VmPolicy.Builder()
                    .detectAll()
                    .penaltyDeath()
                    .penaltyLog()
                    .build());
        }

        mObjectGraph = ObjectGraph.create(
                new ContextProvider(this),
                new ApplicationModule(),
                new ViewUtilProvider(),
                new TaskProvider(),
                new InjectorModule(this),
                new ReceiverProvider()
        );

        mObjectGraph.inject(this);
    }
```

```

    public MainController getMainController() {
        return mMainController;
    }

    public ObjectGraph getObjectGraph() {
        return mObjectGraph;
    }

    @Override
    public void inject(Object object) {
        mObjectGraph.inject(object);
    }
}
```

当@Inject 和@Provides 注解的类构建了一个对象之间相互依赖的图表关联时，通过 ObjectGraph.create()方法获取这个对象图表，其中参数为所有的 Module。

ObjectGraph 的主要方法如下。

- create(Object... modules)：这是个静态的构造方法，用于返回一个 ObjectGraph 的实例，是使用 Dagger 调用的第一个方法。参数为 ModuleClass 对象，方法的作用是根据 ModuleClass 构建一个依赖关系图。此方法的实现会直接调用 DaggerObjectGraph.makeGraph(null, new FailoverLoader(), modules)返回一个 DaggerObjectGraph 对象。
- inject(T instance)：抽象方法，表示向某个 Host 对象中注入依赖。
- injectStatics()：抽象方法，表示向 ObjectGraph 中相关的 Host 注入静态属性。
- get(Class type)：抽象方法，表示得到某个对象的实例，多用于得到依赖的实例。
- plus(Object... modules)：抽象方法，表示返回一个新的包含当前 ObjectGraph 中所有 Binding 的 ObjectGraph。
- validate()：抽象方法，表示对当前的 ObjectGraph 做检查。

(6) 延迟注入。

某些情况下需要延迟初始化一个对象。对任意的对象 T 来说，可以使用 Lazy 实现延迟初始化。Lazy 只有当调用 Lazy 的 get()方法时，才会初始化 T 对象。如果 T 是一个单独的对象，Lazy 也将使用同一个对象进行注入操作。否则，每次注入都将生成自己的 Lazy 对象。当然，任何随后调用 Lazy.get()方法的操作将返回之前构建好的 T 对象。

例如，在 BaseMovieRunnable 中延迟注入的实现代码如下：

```
public abstract class BaseMovieRunnable<R> extends NetworkCallRunnable<R> {

    ...
    @Inject Lazy<Tmdb> mLazyTmdbClient;
    @Inject Lazy<Trakt> mLazyTraktClient;
    @Inject Lazy<AsyncDatabaseHelper> mDbHelper;
    @Inject Lazy<TraktMovieEntityMapper> mLazyTraktMovieEntityMapper;
```

```
    @Inject Lazy<TmdbMovieEntityMapper> mLazyTmdbMovieEntityMapper;
    @Inject Lazy<TmdbCastEntityMapper> mLazyTmdbCastEntityMapper;
    @Inject Lazy<TmdbCrewEntityMapper> mLazyTmdbCrewEntityMapper;
    @Inject Lazy<TmdbPersonEntityMapper> mLazyTmdbPersonEntityMapper;
    @Inject Lazy<Bus> mEventBus;
    @Inject Lazy<CountryProvider> mCountryProvider;

    public BaseMovieRunnable(int callingId) {
        mCallingId = callingId;
    }
}
```

9.3 网络接口设计与数据解析

Philm 使用 Square Inc. 提供的 OkHttp、Retrofit 与 GSON 简单、快速地集成 REST API,为项目的网络功能实现提供了良好的设计模式。

9.3.1 网络接口设计

本节以获取图 9-1 所显示的当前流行电影信息为例,分析基于 MVP 设计模式和 REST 软件架构风格的网络接口设计。

1. MVP 框架流程

PopularMoviesFragment 继承自 BaseMovieControllerListFragment(具体的类层次关系参见 9.4.3 小节),当该 Fragment 被渲染时会调用到 BaseMovieControllerListFragment 的 onResume()方法,在 onResume()方法中会继续调用 MovieController 的 attachUi()方法(由于 MovieController 继承自 BaseUiController,本质上调用的是 BaseUiController 的 attachUi()方法,具体代码参见 9.2.1 小节)。

在 attachUi() 方法中调用 onUiAttached(ui) 方法和 populateUi(ui) 方法。onUiAttached()方法在 MovieController 中进行了重新实现,核心代码如下:

```
@Override
protected void onUiAttached(final MovieUi ui) {
    final MovieQueryType queryType = ui.getMovieQueryType();

    if (queryType.requireLogin() && !isLoggedIn()) {
        return;
    }

    String title = null;
    String subtitle = null;

    final int callingId = getId(ui);

```

```
    switch (queryType) {
        case TRENDING:
            fetchTrendingIfNeeded(callingId);
            break;
        case POPULAR:
            fetchPopularIfNeeded(callingId);
            break;
        ...
    }

    final Display display = getDisplay();
    if (display != null) {
        if (!ui.isModal()) {
            display.showUpNavigation(queryType != null &&
                    queryType.showUpNavigation());
            display.setColorScheme(getColorSchemeForUi(ui));
        }
        display.setActionBarSubtitle(subtitle);
    }
}
```

在 PopularMoviesFragment 中，getMovieQueryType()方法设置 MovieController 的 MovieQueryType 为 POPULAR，因此，在 POPULAR 分支调用 fetchPopularIfNeeded()方法，代码如下：

```
private void fetchPopularIfNeeded(final int callingId) {
    MoviesState.MoviePaginatedResult popular = mMoviesState.getPopular();
    if (popular == null || PhilmCollections.isEmpty(popular.items)) {
        fetchPopular(callingId, TMDB_FIRST_PAGE);
    }
}
```

如果 mMoviesState 中 popular 数据为空，将会执行 fetchPopular()方法，代码如下：

```
private void fetchPopular(final int callingId, final int page) {
    executeTask(new FetchTmdbPopularRunnable(callingId, page));
}
```

其中，executeTask()方法的代码如下：

```
private <R> void executeTask(BaseMovieRunnable<R> task) {
    mInjector.inject(task);
    mExecutor.execute(task);
}
```

mExecutor 是在创建 MovieController 对象时注入的，本质上这个对象是在 app. philm. in. modules. library. UtilProvider 中提供的，代码如下：

```
@Provides @Singleton @GeneralPurpose
public BackgroundExecutor provideMultiThreadExecutor() {
    final int numberCores = Runtime.getRuntime().availableProcessors();
    return new PhilmBackgroundExecutor(
            Executors.newFixedThreadPool(numberCores * 2 + 1));
}
```

所以调用 MovieController 的 executeTask()方法中的 mExecutor.execute(task)，本质上是调用 PhilmBackgroundExecutor 的 execute()方法，代码如下：

```
@Override
public <R> void execute(NetworkCallRunnable<R> runnable) {
    mExecutorService.execute(new TraktNetworkRunner<>(runnable));
}
```

可以看到 FetchTmdbPopularRunnable 对象被传入 TraktNetworkRunner，其中的 run()方法代码如下：

```
01 @Override
02 public final void run() {
03     android.os.Process.setThreadPriority(
04             Process.THREAD_PRIORITY_BACKGROUND);
05
06     sHandler.post(new Runnable() {
07         @Override
08         public void run() {
09             mBackgroundRunnable.onPreTraktCall();
10         }
11     });
12
13     R result = null;
14     RetrofitError retrofitError = null;
15
16     try {
17         result = mBackgroundRunnable.doBackgroundCall();
18     } catch (RetrofitError re) {
19         retrofitError = re;
20         if (Constants.DEBUG) {
21             Log.d(((Object) this).getClass().getSimpleName(),
22                     "Error while completing network call", re);
23         }
24     }
25
26     sHandler.post(new ResultCallback(result, retrofitError));
27 }
```

其中，17 行调用的 doBackgroundCall()方法是 FetchTmdbPopularRunnable 提供的，代码如下：

```
01  @Override
02  public MovieResultsPage doBackgroundCall() throws RetrofitError {
03      return getTmdbClient().moviesService().popular(
04              getPage(),
05              getCountryProvider().getTwoLetterLanguageCode());
06  }
```

03 行的 getTmdbClient()方法返回一个 PhilmTmdb 对象。在 NetworkProvider 中，provideTmdbClient()方法的代码如下：

```
@Provides @Singleton
public Tmdb provideTmdbClient(@CacheDirectory File cacheLocation) {
    Tmdb tmdb = new PhilmTmdb(cacheLocation);
    tmdb.setApiKey(Constants.TMDB_API_KEY);
    tmdb.setIsDebug(Constants.DEBUG_NETWORK);
    return tmdb;
}
```

这是一种通过多态的原理实现的单实例模式。获取的 PhilmTmdb 的 moviesService()方法代码如下：

```
public MoviesService moviesService() {
    return (MoviesService)this.getRestAdapter().create(MoviesService.class);
}
```

这里的 getRestAdapter()和 MoviesService 非常重要，下面将分别介绍。

2. 基于 OkHttp 网络连接

OkHttp[①] 是一个高效的 HTTP 库，具有以下特点：

- 支持 SPDY(Google 开发的基于 TCP 的应用层协议，用于最小化网络延迟，提升网络速度，优化用户的网络使用体验)，共享同一个 Socket 来处理同一个服务器的所有请求。
- 如果 SPDY 不可用，则通过连接池来减少请求延时。
- 无缝地支持 GZIP 来减少数据流量。
- 缓存响应数据来减少重复的网络请求。

下面分析一下前面所述的 moviesService()方法中的 getRestAdapter()在 OkHttp 方面的应用。getRestAdapter()代码如下：

```
protected RestAdapter getRestAdapter() {
    if(this.restAdapter == null) {
        Builder builder = this.newRestAdapterBuilder();
        builder.setEndpoint("https://api.themoviedb.org/3");
        builder.setConverter(new GsonConverter(
```

① OkHttp 官网：http://square.github.io/okhttp/。

```
06                 TmdbHelper.getGsonBuilder().create()));
07         builder.setRequestInterceptor(new RequestInterceptor() {
08             public void intercept(RequestFacade requestFacade) {
09                 requestFacade.addQueryParam("api_key", Tmdb.this.apiKey);
10             }
11         });
12         if(this.isDebug) {
13             builder.setLogLevel(LogLevel.FULL);
14         }
15
16         this.restAdapter = builder.build();
17     }
18
19     return this.restAdapter;
20 }
```

其中 03 行的 newRestAdapterBuilder()在 Tmdb 的 PhilmTmdb 子类中的实现代码如下：

```
01 @Override
02 protected RestAdapter.Builder newRestAdapterBuilder() {
03     RestAdapter.Builder b = super.newRestAdapterBuilder();
04
05     if (mCacheLocation != null) {
06         OkHttpClient client = new OkHttpClient();
07
08         try {
09             File cacheDir = new File(mCacheLocation,
10                     UUID.randomUUID().toString());
11             Cache cache = new Cache(cacheDir, 1024);
12             client.setCache(cache);
13         } catch (IOException e) {
14             Log.e(TAG, "Could not use OkHttp Cache", e);
15         }
16
17         client.setConnectTimeout(Constants.CONNECT_TIMEOUT_MILLIS,
18                 TimeUnit.MILLISECONDS);
19         client.setReadTimeout(Constants.READ_TIMEOUT_MILLIS,
20                 TimeUnit.MILLISECONDS);
21
22         b.setClient(new OkClient(client));
23     }
24
25     return b;
26 }
```

06 行的 OkHttpClient 为 HTTP Client 配置包括代理设置、超时设置、缓存设置等参数。OkHttp 官方文档并不建议创建多个 OkHttpClient，如果需要，可以使用 clone()方法创建一个对象副本，再进行自定义。

3. 基于 Retrofit 的 REST 接口封装

Retrofit[1] 与 Java 领域的 ORM(用于实现面向对象编程语言里不同类型系统的数据之间转换的一种编程技术)概念类似，ORM 把结构化数据转换为 Java 对象，而 Retrofit 把 REST API 返回的数据转化为 Java 对象以方便操作。Retrofit 支持 URL 参数替换和查询参数，支持 Multipart 请求和文件上传，同时还封装了网络代码的调用。

Retrofit 使用的基本步骤如下：

① 定义接口，参数声明，Url 通过 Annotation 注解。

moviesService()方法相关的 MoviesService 核心代码如下：

```
public interface MoviesService {
    ...

    @GET("/movie/popular")
    MovieResultsPage popular(@Query("page") Integer var1, @Query("language")
            String var2);

    @GET("/movie/{id}/images")
    Images images(@Path("id") int var1, @Query("language") String var2);
}
```

定义上面的是一个 REST API 接口，该接口定义了一个 popular()方法，该方法会通过 HTTP GET 请求去访问服务器的/movie/popular 路径并把返回的结果封装为 MovieResultsPage Java 对象后返回。

Retrofit 的 Annotation 包含请求方法相关的@GET、@POST、@HEAD、@PUT、@DELETA、@PATCH 和参数相关的@Path、@Field、@Multipart 等。

② 通过 RestAdapter 生成一个接口的实现类(动态代理)，而这个实现类是通过 RestAdapter.create()方法返回的。例如：

```
public MoviesService moviesService() {
    return (MoviesService)this.getRestAdapter().create(MoviesService.class);
}
```

③ 调用接口请求数据。例如前面描述的 FetchTmdbPopularRunnable 中的 doBackgroundCall()方法。

9.3.2 数据解析与显示

1. 解析数据

当 TraktNetworkRunner 执行完后会发出 sHandler.post(new ResultCallback(result, retrofitError))请求，里面包含了请求的反馈结果，用于修改 UI 的数据。其中 ResultCallback()

① Retrofit 官网：http://square.github.io/retrofit/。

方法代码如下：

```
private class ResultCallback implements Runnable {
    private final R mResult;
    private final RetrofitError mRetrofitError;

    private ResultCallback(R result, RetrofitError retrofitError) {
        mResult = result;
        mRetrofitError = retrofitError;
    }

    @Override
    public void run() {
        if (mResult != null) {
            mBackgroundRunnable.onSuccess(mResult);
        } else if (mRetrofitError != null) {
            mBackgroundRunnable.onError(mRetrofitError);
        }
        mBackgroundRunnable.onFinished();
    }
}
```

此时，执行 FetchTmdbPopularRunnable 的祖父类 BaseTmdbPaginatedRunnable 中的 onSuccess()方法，代码如下：

```
01 @Override
02 public final void onSuccess(TR result) {
03     if (result != null) {
04         R paginatedResult = getResultFromState();
05
06         if (paginatedResult == null) {
07             paginatedResult = createPaginatedResult();
08             paginatedResult.items = new ArrayList<>();
09         }
10
11         updatePaginatedResult(paginatedResult, result);
12         updateState(paginatedResult);
13     }
14 }
```

11～12 行分别执行下面的两个回调方法来保存获取的数据并显示在用户界面中。

```
@Override
protected MoviesState.MoviePaginatedResult getResultFromState() {
    return mMoviesState.getPopular();
}

@Override
protected void updateState(MoviesState.MoviePaginatedResult result) {
```

```
    mMoviesState.setPopular(result);
}
```

同时，在 FetchTmdbPopularRunnable 的父类 BaseTmdbPaginatedMovieRunnable 中执行回调方法 updatePaginatedResult()，代码如下：

```
01 @Override
02 protected void updatePaginatedResult(
03         MoviesState.MoviePaginatedResult result,
04         MovieResultsPage tmdbResult) {
05     result.items.addAll(getTmdbMovieEntityMapper()
06             .mapAll(tmdbResult.results));
07
08     result.page = tmdbResult.page;
09     if (tmdbResult.total_pages != null) {
10         result.totalPages = tmdbResult.total_pages;
11     }
12 }
```

05 行的 getTmdbMovieEntityMapper()方法返回 TmdbMovieEntityMapper 对象，该对象是 BaseEntityMapper 的子类，其 mapAll()方法的代码如下：

```
01 public List<R> mapAll(List<T> entities) {
02     final ArrayList<R> movies = new ArrayList<>(entities.size());
03     for (T entity : entities) {
04         movies.add(map(entity));
05     }
06     return movies;
07 }
```

04 行的 map()方法通过多态的原理在 TmdbMovieEntityMapper 中实现，代码如下：

```
@Override
public PhilmMovie map(Movie entity) {
    PhilmMovie movie = getEntity(String.valueOf(entity.id));

    if (movie == null && entity.imdb_id != null) {
        movie = getEntity(entity.imdb_id);
    }

    if (movie == null) {
        //No movie, so create one
        movie = new PhilmMovie();
    }
    //We already have a movie, so just update it wrapped value
    movie.setFromMovie(entity);
    putEntity(movie);
```

```

    return movie;
}
```

2. 显示数据

在 MVP 中,Model 的数据被更新后,需要通过 Presenter(MovieController)来更新 View(Model 与 View 不能直接通信)。Philm 通知 Presenter 是通过时间总线来实现的(使用的是 Square 的 otto 消息库)。

当返回数据解析完毕并把结果保存在 mMoviesState 中后,调用其 setPopular()方法发布消息,代码如下:

```
@Override
public void setPopular(MoviePaginatedResult items) {
    mPopular = items;
    mEventBus.post(new PopularChangedEvent());
}
```

其中,发布事件的语句 mEventBus. post(new PopularChangedEvent())使用的是 otto 库的消息队列。在 MovieController 中,通过@Subscribe 被监听,代码如下:

```
@Subscribe
public void onPopularChanged(MoviesState.PopularChangedEvent event) {
    populateUiFromQueryType(MovieQueryType.POPULAR);
}

private final void populateUiFromQueryType(MovieQueryType queryType) {
    MovieUi ui = findUiFromQueryType(queryType);
    if (ui != null) {
        populateUi(ui);
    }
}

private MovieUi findUiFromQueryType(MovieQueryType queryType) {
    for (MovieUi ui : getUis()) {
        if (ui.getMovieQueryType() == queryType) {
            return ui;
        }
    }
    return null;
}
```

执行变更视图的类型是 MovieQueryType. POPULAR(在 PopularMoviesFragment 的 onResume()中被注册到 MovieController 中)。

PopularMoviesFragment 实现了 MovieController. MovieListUi 接口,因此,populateUi()执行的本质是 MovieController 中的 populateMovieListUi()方法,核心代码如下:

```
01 private void populateMovieListUi(MovieListUi ui) {
02     final MovieQueryType queryType = ui.getMovieQueryType();
03     ...
04     if (items == null) {
05         ui.setItems(null);
06     } else if (PhilmCollections.isEmpty(sections)) {
07         ui.setItems(createListItemList(items));
08 
09         if (isLoggedIn()) {
10             ui.allowedBatchOperations(MovieOperation.MARK_SEEN,
11                     MovieOperation.ADD_TO_COLLECTION,
12                     MovieOperation.ADD_TO_WATCHLIST);
13         } else {
14             ui.disableBatchOperations();
15         }
16     } else {
17         ui.setItems(createSectionedListItemList(items, sections,
18                 sectionProcessingOrder));
19     }
20 }
```

其中 17～18 行的 setItems()方法代码如下：

```
@Override
public void setItems(List<ListItem<PhilmMovie>> items) {
    mMovieGridAdapter.setItems(items);
    moveListViewToSavedPositions();
}
```

而 mMovieGridAdapter 的 setItems()方法代码如下：

```
public void setItems(List<ListItem<PhilmMovie>> items) {
    if (!Objects.equal(items, mItems)) {
        mItems = items;
        notifyDataSetChanged();
    }
}
```

可以看到，到这里在 UI 中就获得了数据，然后可以根据 URL 下载电影海报等信息。

9.4　UI 设计

Philm 的 UI 定制了 MovieDetailCardLayout、MovieDetailInfoLayout、SlidingTabLayout 和 ViewRecycler 等大量的 View 及其管理工具，并结合 Material Design 设计出体验

良好的应用。

9.4.1 Material Design

虽然 Philm 没有明确提出基于 Google 最新的 Material Design 风格,但是通过其样式设计和布局设计的特点可以看出,Philm 采用了大量的 Material Design 元素。

1. 设计原则

Material Design 设计的基本原则如下。

(1) 实体感就是隐喻

通过构建系统化的动效和空间合理化利用,并将两个理念合二为一,构成了实体隐喻。实体的表面和边缘提供基于真实效果的视觉体验,熟悉的触感让用户可以快速地理解和认知。实体的多样性可以让应用呈现出更多反映真实世界的设计效果,但同时又绝不会脱离客观的物理规律。

在电影详情界面,界面的首页显示了图 9-17 中的大背景海报和大字体电影名称,随着手指的向上滑动,电影海报逐渐缩减成如图 9-18 所示的剪裁的海报背景,同时,电影的名称也逐渐缩小到界面标题的位置。

图 9-17 电影详情界面首页

图 9-18 向上滑动后的首页

(2) 鲜明、形象、深思熟虑

新的视觉语言,在基本元素的处理上,借鉴了传统的印刷设计——排版、网格、空间、比例、配色、图像使用——这些基础的平面设计规范。在这些设计基础上下功夫,不但可以愉悦用户,而且能够构建出视觉层级、视觉意义以及视觉聚焦。精心选择色彩、图像,选择合乎比例的字体、留白,力求构建出鲜明、形象的用户界面,让用户沉浸其中。

在 Philm 中,主界面(MainActivity)包含的 Fragment 及 ViewPage 采用了唯一主色调风格,这样能够很好地表达界面层次、重要信息,并且能展现良好的视觉效果。现在越来越多唯一主色调风格的设计,多采用简单的色阶、配套灰阶来展现信息层次,但是绝不采用更多的颜色。

(3) 有意义的动画效果

动画效果(简称动效)可以有效地暗示、指引用户。动效的设计要根据用户行为而定,能够改变整体设计的触感。动效应当在独立的场景中呈现。通过动效，让物体的变化以更连续、更平滑的方式呈现给用户，让用户能够充分知晓所发生的变化。

2. 布局原则

Material Design 使用的基本工具来源于印刷设计，例如通用于所有界面的基准线和栅格。布局排版能够按比例横跨不同尺寸的屏幕，促进 UI 开发和从根本上帮助设计师做可扩展的应用。

(1) 电影详情界面

电影详情界面如图 9-3 所示，该界面基本遵循了 Material Design 中关于水平边距设计的准则(图 9-19)。

其中，❶、❷、❸、❹处的尺寸分别是 24dp、56dp、8dp 和 72dp。

提示：dp 是 Android 中非文字的尺寸单位，是 density-independent pixels(密度独立像素)的缩写。

(2) 搜索界面

搜索界面如图 9-11 所示，该界面基本遵循了图 9-20 所示的 Material Design 设计原则。其中，❶、❷、❸、❹、❺处的尺寸分别是 24dp、56dp、72dp、48dp 和 8dp。

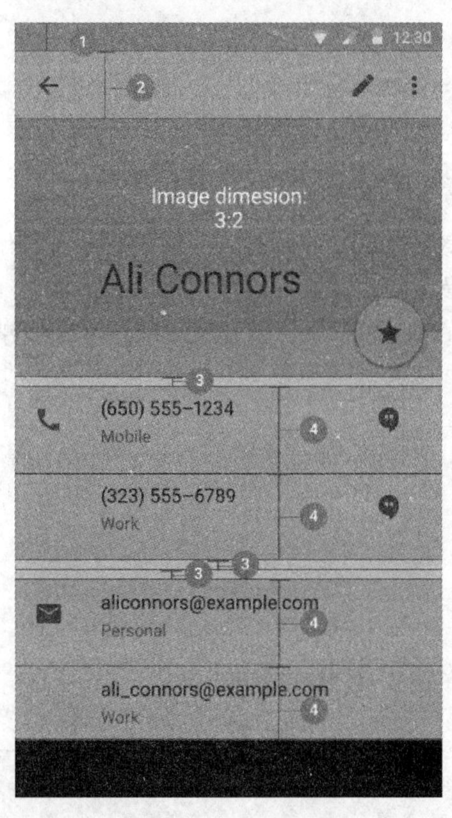

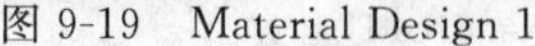
图 9-19　Material Design 1

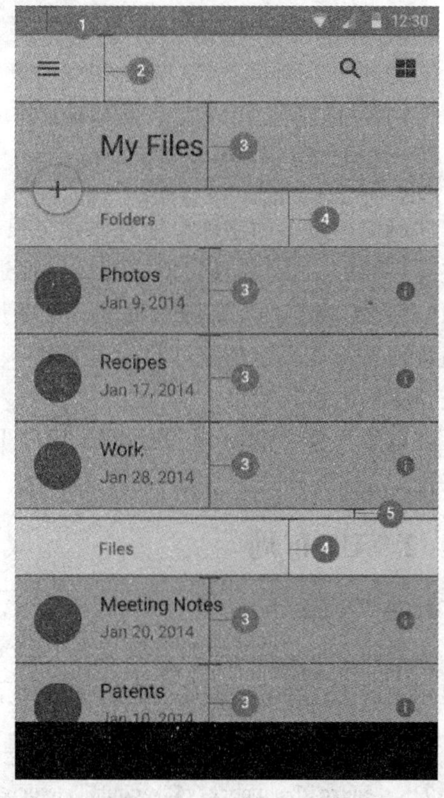

图 9-20　Material Design 2

3. 字体原则

自从 Ice Cream Sandwich 发布以来，Roboto 都是 Android 系统的默认字体集，如图 9-21 所示。

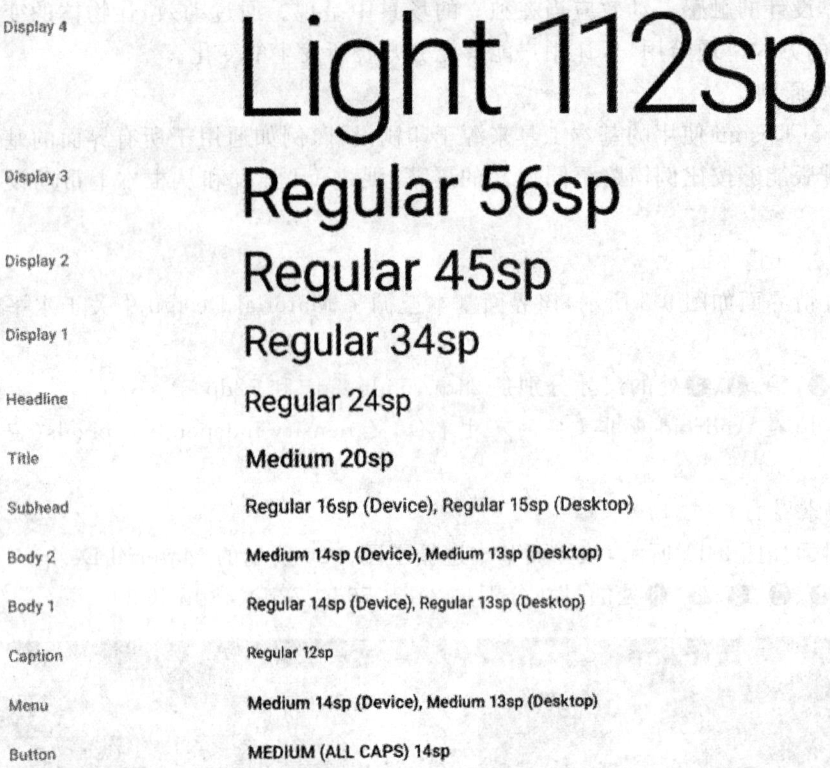

Display 4	Light 112sp
Display 3	Regular 56sp
Display 2	Regular 45sp
Display 1	Regular 34sp
Headline	Regular 24sp
Title	Medium 20sp
Subhead	Regular 16sp (Device), Regular 15sp (Desktop)
Body 2	Medium 14sp (Device), Medium 13sp (Desktop)
Body 1	Regular 14sp (Device), Regular 13sp (Desktop)
Caption	Regular 12sp
Menu	Medium 14sp (Device), Medium 13sp (Desktop)
Button	MEDIUM (ALL CAPS) 14sp

图 9-21　Roboto 字体样式

同时使用过多的字体尺寸和样式会很轻易地毁掉布局。字体排版的缩放是包含了有限个字体尺寸的集合,并且它们能够良好地适应布局结构。最基本的样式集合就是基于 12、14、16、20 和 34 号的字体排版缩放。这些尺寸和样式在经典应用场合中让内容密度和阅读舒适度取得平衡。字体尺寸是通过 SP(scaleable pixels,可缩放像素数)指定的,让大尺寸字体获得更好的可接受度。

在电影详情界面中,电影名称采用了 Display1 的样式,评分采用 Headlne 样式,电影简介采用 Caption 样式。

9.4.2　UI 布局

1. 主界面设计

主界面 MainActivity 的布局是一个 DrawerLayout,由 MenuFragment(SideMenuFragment)和显示内容 Fragment 组成。布局文件如下:

```
<?xml version = "1.0" encoding = "utf - 8"?>
< android. support. v4. widget. DrawerLayout
xmlns:android = "http://schemas. android. com/apk/res/android"
    android: id = "@ + id/drawer_layout"
    android: layout_width = "match_parent"
    android: layout_height = "match_parent"
    android: fitsSystemWindows = "true" >
```

```
<include layout = "@layout/activity_no_drawer" />

<fragment
    android:id = "@ + id/fragment_side_menu"
    android:name = "app.philm.in.fragments.SideMenuFragment"
    android:layout_width = "@dimen/drawer_side_menu_width"
    android:layout_height = "match_parent"
    android:layout_gravity = "left" />

</android.support.v4.widget.DrawerLayout>
```

在 Fragment 中以 Tab 页组织 UI 界面，其中嵌入了 Viewpager 来管理多个 Fragment。使用 Tab 将大量关联的数据或者选项划分成更易理解的分组，可以在不需要切换出当前上下文的情况下，有效地进行内容的导航和组织。

该界面的设计原则如下。

- Tab 的宽度为：12dp＋文本宽度＋12dp。
- 激活的 Tab 指示器高度为 2dp。
- 文本属性为 14sp、Roboto 和 Medium。
- 文本在 Tab 中居中。
- 激活的文字颜色为＃fff 或颜色选择中的次要颜色。
- 不可用的文字颜色为 60％的＃fff。

2. 电影详情界面设计

电影详情界面以卡片式设计为主，其中显示电影详情的布局文件如下：

```
<?xml version = "1.0" encoding = "utf - 8"?>
<FrameLayout xmlns:android = "http://schemas.android.com/apk/res/android"
        xmlns:app = "http://schemas.android.com/apk/res - auto"
        android:layout_height = "match_parent"
        android:layout_width = "match_parent">

    <include layout = "@layout/include_fragment_detail_list"/>

    <app.philm.in.view.CollapsingTitleLayout
        android:id = "@ + id/backdrop_toolbar"
        android:layout_width = "match_parent"
        android:layout_height = "wrap_content"
        android:elevation = "@dimen/movie_detail_pinned_elevation"
        android:textAppearance = "@style/TextAppearance.AppCompat.
                Widget.ActionBar.Title.Inverse"
        app:expandedTextSize = "56dp"
        app:expandedMargin = "16dp">

        <app.philm.in.view.BackdropImageView
            android:id = "@ + id/imageview_fanart"
```

```
                android:layout_width = "match_parent"
                android:layout_height = "@dimen/movie_detail_fanart_height"
                android:scaleType = "centerCrop"
                android:background = "?attr/colorPrimary" />

        <android.support.v7.widget.Toolbar
                android:id = "@+id/toolbar"
                android:layout_height = "?attr/actionBarSize"
                android:layout_width = "match_parent"
                style = "@style/Widget.Philm.Toolbar.Transparent" />

    </app.philm.in.view.CollapsingTitleLayout>

</FrameLayout>
```

卡片是一种采用较多的设计语言形式，卡片式设计在栅格的基础上更进了一步，将整个页面的内容切割为许多个区域，不仅能给人很好的视觉一致性，而且更易于实现设计上的迭代。Google 的移动端产品设计已经全面卡片化了，甚至 Web 端也沿用了这种统一的设计语言。

对于移动设备碎片化的屏幕来说，卡片是完美的设计形式。在手机上，卡片通常以垂直方式展现。电影详情界面使用卡片式列表框架展示信息，卡片流突出信息本身，用大图和标题文字吸引用户，强化了无尽浏览的体验。Philm 中，不同的卡片都遵循在一个统一宽度和样式的卡片内，进行发挥和设计，既保证了卡片和卡片之间的独立性，又保证了服务和服务的统一化设计。

3. 登录界面设计

登录界面以扁平化的设计风格展示界面信息，布局核心代码如下：

```
<?xml version = "1.0" encoding = "utf-8"?>
<LinearLayout xmlns:android = "http://schemas.android.com/apk/res/android"
    xmlns:philm = "http://schemas.android.com/apk/res-auto"
    android:layout_width = "match_parent"
    android:layout_height = "match_parent"
    android:orientation = "vertical" >

    <ScrollView
        android:layout_width = "match_parent"
        android:layout_height = "0px"
        android:layout_weight = "1" >

        <LinearLayout
            ...
            android:orientation = "vertical"
            android:padding = "@dimen/spacing_major" >

            <app.philm.in.view.FontTextView ... />

```

```
            <app.philm.in.view.FloatLabelLayout ... >
                <EditText ... />
            </app.philm.in.view.FloatLabelLayout>
            <app.philm.in.view.FloatLabelLayout ... >
                <EditText ... />
            </app.philm.in.view.FloatLabelLayout>
            <app.philm.in.view.FloatLabelLayout ... >
                <AutoCompleteTextView ... />
            </app.philm.in.view.FloatLabelLayout>
            <RadioGroup
                ...
            </RadioGroup>
        </LinearLayout>
    </ScrollView>
    <View
        android:layout_width = "match_parent"
        android:layout_height = "1dp"
        android:background = "?android:attr/dividerVertical" />
    <LinearLayout
        ... >
        <Button ... />
    </LinearLayout>
</LinearLayout>
```

扁平化设计是一种设计风格术语，它抛弃任何能使得作品凸显 3D 效果的特性。通俗地说，设计中不使用透视、纹理、阴影等效果。

扁平化设计通常采用许多简单的用户界面元素，诸如按钮或者图标之类。设计师通常坚持使用简单的外形(矩形或者圆形)，并且尽量突出外形，这些元素一律为直角(极个别的为圆角)。这些用户界面元素方便用户点击，能极大地减少用户学习新交互方式的成本，因为用户凭经验就能大概知道每个按钮的作用。

扁平化设计在整体上趋近极简主义设计理念。设计师要尽量简化自己的设计方案，避免不必要的元素出现在设计中。

9.4.3 Fragment 设计

Fragment 是 Android API 11 引入的概念，Fragment 为开发者和设计师提供了一种全新的方法，让他们设计的应用变得有弹性、可堆叠，从而适应不同设备的屏幕规格。屏幕组件可以自由拉伸、堆叠、缩放和隐藏,以支持更动态、更灵活的界面设计。

1. 类层次结构

下面以图 9-1 显示的当前流行电影的 Fragment——PopularMoviesFragment 为例说明。通过代码分析可以看出,PopularMoviesFragment 存在图 9-22 所示的类继承关系。

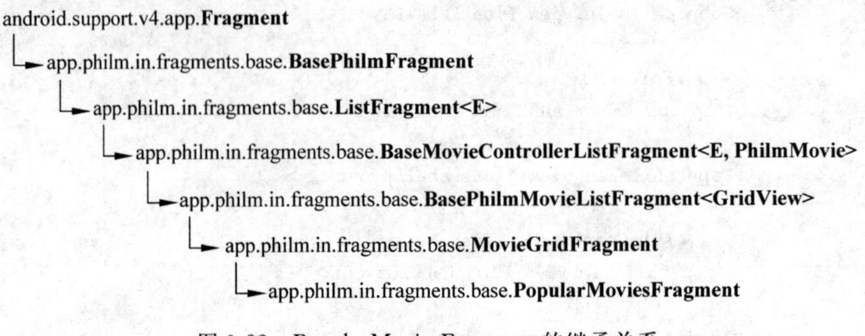

图 9-22 PopularMoviesFragment 的继承关系

(1) PopularMoviesFragment

实现 MovieController. SubUi 接口,并实现如下回调方法:

```
@Override
public MovieController.MovieQueryType getMovieQueryType() {
    return MovieController.MovieQueryType.POPULAR;
}
```

该回调方法的主要作用是告诉控制器当前界面需要显示的数据是流行电影信息。

(2) MovieGridFragment

MovieGridFragment 的核心任务包括两点。

① 创建并初始化 MovieGridAdapter 适配器,然后将数据显示在内嵌的 GridView 控件上。代码如下:

```
@Override
public void onCreate(Bundle savedInstanceState) {
    super.onCreate(savedInstanceState);

    mMovieGridAdapter = new MovieGridAdapter(getActivity());
    setListAdapter(mMovieGridAdapter);
}
```

② 设置 GridView 每个 Item 的单击事件,即在界面上单击每个电影海报,可以显示电影详情的事件处理情况。代码如下:

```
@Override
public void onListItemClick(GridView l, View v, int position, long id) {
    if (hasCallbacks()) {
        ListItem<PhilmMovie> item = (ListItem<PhilmMovie>) l
                .getItemAtPosition(position);
```

```
        if (item.getListType() == ListItem.TYPE_ITEM) {
            getCallbacks().showMovieDetail(item.getListItem(),
                    ActivityTransitions.scaleUpAnimation(v));
        }
    }
}
```

(3) BasePhilmMovieListFragment<GridView>

BasePhilmMovieListFragment 实现 MovieController. MovieListUi 和 AbsListView. OnScrollListener 接口。在功能上，主要完成如下任务：

① 通过 onCreateOptionsMenu()、onPrepareOptionsMenu()和 onOptionsItemSelected()三个回调方法创建选项菜单。

② 设置长按 GridView 中的 Item，即长按电影海报时的动作，显示效果如图 9-23 所示。

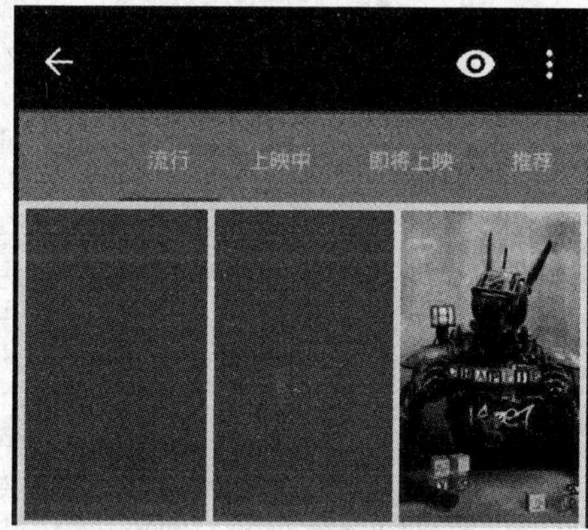

图 9-23 长按海报动作栏的设计

长按响应动作栏显示后，可以单击多个电影海报，每个被选中的海报以矩形填充方块显示，以支持批量处理。核心代码如下：

```
private class MovieMultiChoiceListener implements
        AbsListView.MultiChoiceModeListener {

    @Override
    public void onItemCheckedStateChanged(ActionMode mode, int position,
            long id, boolean checked) {
        //NO - OP
    }

    @Override
    public boolean onCreateActionMode(ActionMode mode, Menu menu) {
        mode.getMenuInflater().inflate(R.menu.cab_movies, menu);

        for (int i = 0, z = mBatchOperations.length; i < z; i++) {
            switch (mBatchOperations[i]) {
                case ADD_TO_COLLECTION:
                    menu.findItem(R.id.menu_add_collection).setVisible(true);
                    break;
                case ADD_TO_WATCHLIST:
                    menu.findItem(R.id.menu_add_watchlist).setVisible(true);
                    break;
                case MARK_SEEN:
                    menu.findItem(R.id.menu_mark_seen).setVisible(true);
                    break;
```

```
            }
        }

        return true;
    }

    @Override
    public boolean onPrepareActionMode(ActionMode mode, Menu menu) {
        return false;
    }

    @Override
    public boolean onActionItemClicked(ActionMode mode, MenuItem item) {
        if (hasCallbacks()) {
            final E listView = getListView();
            final SparseBooleanArray checkedItems =
                    listView.getCheckedItemPositions();
            final ArrayList<PhilmMovie> movies = new
                    ArrayList<>(checkedItems.size());
            for (int i = 0, z = checkedItems.size(); i < z; i++) {
                if (checkedItems.valueAt(i)) {
                    final int index = checkedItems.keyAt(i);
                    ListItem<PhilmMovie> listItem = (ListItem<PhilmMovie>)
                            listView.getItemAtPosition(index);
                    if (listItem.getListType() == ListItem.TYPE_ITEM) {
                        movies.add(listItem.getListItem());
                    }
                }
            }

            switch (item.getItemId()) {
                case R.id.menu_mark_seen:
                    getCallbacks().setMoviesSeen(movies, true);
                    return true;
                case R.id.menu_add_collection:
                    getCallbacks().setMoviesInCollection(movies, true);
                    return true;
                case R.id.menu_add_watchlist:
                    getCallbacks().setMoviesInWatchlist(movies, true);
                    return true;
            }

        }
        return false;
    }

    @Override
    public void onDestroyActionMode(ActionMode mode) {
        //NO-OP
    }
}
```

(4) BaseMovieControllerListFragment<E,PhilmMovie>

BaseMovieControllerListFragment 实现 MovieController. BaseMovieListUi<T>和 AbsListView. OnScrollListener 接口。在功能上,主要完成如下任务。

① 在 onResume()中调用 getController(). attachUi(this)进行 UI 的渲染,这个 UI 对应的 Controller 为 MovieController,在 MovieController 中先是执行 BaseUiController 中的 attachUi,然后紧接着执行 onUiAttached,这个方法在 MovieController 中实现。

② 保存当前电影海报的浏览位置,代码如下:

```
private void saveListViewPosition() {
    E listView = getListView();

    mFirstVisiblePosition = listView.getFirstVisiblePosition();

    if (mFirstVisiblePosition != AdapterView.INVALID_POSITION &&
            listView.getChildCount() > 0) {
        mFirstVisiblePositionTop = listView.getChildAt(0).getTop();
    }
}
```

③ 实现界面的上拉刷新,即加载更多的电影信息,核心代码如下:

```
@Override
public final void onScrollStateChanged(AbsListView view, int scrollState) {
    if (scrollState == AbsListView.OnScrollListener.SCROLL_STATE_IDLE &&
            mLoadMoreIsAtBottom) {
        if (onScrolledToBottom()) {
            mLoadMoreRequestedItemCount = view.getCount();
            mLoadMoreIsAtBottom = false;
        }
    }
}

@Override
public final void onScroll(AbsListView view, int firstVisibleItem,
        int visibleItemCount, int totalItemCount) {
    mLoadMoreIsAtBottom = totalItemCount > mLoadMoreRequestedItemCount
            && firstVisibleItem + visibleItemCount == totalItemCount;
}

protected boolean onScrolledToBottom() {
    if (Constants.DEBUG) {
        Log.d(LOG_TAG, "onScrolledToBottom");
    }
    if (hasCallbacks()) {
        getCallbacks().onScrolledToBottom();
        return true;
    }
    return false;
}
```

(5) ListFragment<E>

ListFragment 主要是初始化 UI 布局以及涉及的相关动画。

(6) BasePhilmFragment

BasePhilmFragment 主要是设置对动作栏的支持。

2. 数据适配

PopularMoviesFragment 以 GridView 填充用户界面,因此,当从开放接口获取相关数据后,通过继承自 BaseAdapter 的 MovieGridAdapter 适配器将数据适配在 Fragment 中。

MovieGridAdapter 的核心代码如下:

```
public class MovieGridAdapter extends BaseAdapter {

    ...

    public MovieGridAdapter(Activity activity) {
        mActivity = activity;
        mLayoutInflater = mActivity.getLayoutInflater();
    }

    public void setItems(List<ListItem<PhilmMovie>> items) {
        if (!Objects.equal(items, mItems)) {
            mItems = items;
            notifyDataSetChanged();
        }
    }

    @Override
    public int getCount() {
        return mItems != null ? mItems.size() : 0;
    }

    @Override
    public ListItem<PhilmMovie> getItem(int position) {
        return mItems.get(position);
    }

    @Override
    public long getItemId(int position) {
        return position;
    }

    @Override
    public View getView(int position, View convertView, ViewGroup viewGroup) {
        View view = convertView;
        if (view == null) {
            view = mLayoutInflater.inflate(R.layout.item_grid_movie,
                    viewGroup, false);
        }

```

```
        final PhilmMovie movie = getItem(position).getListItem();

        final TextView title = (TextView) view
                .findViewById(R.id.textview_title);
        title.setText(movie.getTitle());
        title.setVisibility(View.VISIBLE);

        final PhilmImageView imageView = (PhilmImageView) view
                .findViewById(R.id.imageview_poster);
        imageView.setAutoFade(false);
        imageView.loadPoster(movie, new PhilmImageView.Listener() {
            @Override
            public void onSuccess(PhilmImageView imageView, Bitmap bitmap) {
                title.setVisibility(View.GONE);
            }

            @Override
            public void onError(PhilmImageView imageView) {
                title.setVisibility(View.VISIBLE);
            }
        });

        return view;
    }
}
```

9.4.4 Activity 实现

Philm 项目共包含 7 个 Activity：MainActivity、MovieActivity、PersonActivity、AccountActivity、AboutActivity、SettingsActivity 和 MovieImagesActivity。

下面主要分析 MainActivity 的设计。

在 AndroidManifest.xml 中，关于 MainActivity 的声明信息如下：

```
01 <activity
02         android:name=".MainActivity"
03         android:label="@string/app_name"
04         android:launchMode="singleTask">
05
06     <intent-filter>
07         <action android:name="android.intent.action.MAIN"/>
08         <category android:name="android.intent.category.LAUNCHER"/>
09     </intent-filter>
10
11 </activity>
```

04 行的 android:launchMode 属性值为 singleTask。设置了 singleTask 启动模式的 Activity 不是在新的任务中启动时，它会在已有的任务中查看是否已经存在相应的 Activity

实例,如果存在,就会把位于这个 Activity 实例上面的 Activity 全部结束,即最终这个 Activity 实例会位于任务的堆栈顶端。因此,保证了 Activity 只实例化一次,由此所开启的 Activity 和本 Activity 位于同一任务中。但是,如果 Activity 已经在其他任务中存在实例,则系统会通过调用其实例的 onNewIntent()方法把 Intent 传给已有实例,而不是再创建一个新实例。

在其他 Activity 的声明中,多有如下信息:

```
<activity
        android:name = ".AboutActivity"
        ...
        android:parentActivityName = ".MainActivity">

    <meta - data
          android:name = "android.support.PARENT_ACTIVITY"
          android:value = ".MainActivity" />

</activity>
```

其中,android:parentActivityName 的作用是给当前 Activity 的界面左上角添加一个返回按钮,android:name="android. support. PARENT_ACTIVITY"属性用于指定:当单击界面左上角的返回按钮时,返回到 android:value 指定的 Activity 界面。

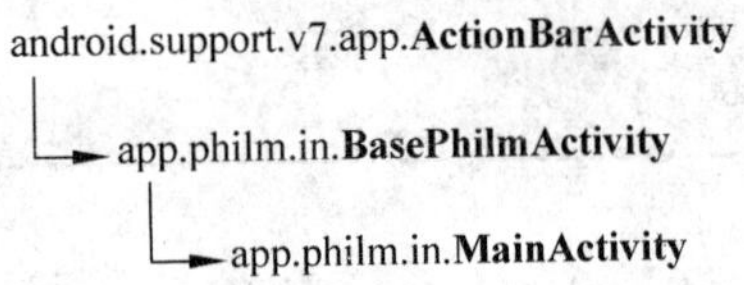

MainActivity 的继承关系如图 9-24 所示。

android.support.v7.app.**ActionBarActivity**
→ app.philm.in.**BasePhilmActivity**
→ app.philm.in.**MainActivity**

图 9-24 MainActivity 的继承关系

(1) MainActivity

MainActivity 主要完成如下工作:

① 重写 onCreateOptionsMenu()回调方法来创建选项菜单。

② 在 onResume()生命周期方法中调用 getMainController(). attachUi(this)方法实现视图渲染;在 onPause()生命周期方法中调用 getMainController(). detachUi(this)方法清空 CallBack。

③ 重写 handleIntent()回调方法来设置侧边菜单。

(2) BasePhilmActivity

BasePhilmActivity 主要完成如下工作:

① 在 onCreate()、onResume()生命周期方法中初始化整个项目的 MainController 和 Display,核心代码如下:

```
@Override
protected void onCreate(Bundle savedInstanceState) {
    ...
    setContentView(getContentViewLayoutId());
    mMainController = PhilmApplication.from(this).getMainController();
    mDisplay = new AndroidDisplay(this, mDrawerLayout);
    handleIntent(getIntent(), getDisplay());
```

```
}

@Override
protected void onResume() {
    super.onResume();
    mMainController.attachDisplay(mDisplay);
    mMainController.setHostCallbacks(this);
    mMainController.init();
}
```

② 重写 onOptionsItemSelected()回调方法响应菜单事件。核心代码如下：

```
@Override
public boolean onOptionsItemSelected(MenuItem item) {
    switch (item.getItemId()) {
        case android.R.id.home:
            if (getMainController().onHomeButtonPressed()) {
                return true;
            }
            if (navigateUp()) {
                return true;
            }
            break;

        ...
    }

    return super.onOptionsItemSelected(item);
}
```

参考文献

[1] Maximiliano Firtman. Programming the Mobile Web[M]. O'Reilly Media,2013.

[2] Joseph Annuzzi. Advanced Android Application Development[M]. 4 版. Addison-Wesley Professional,2014.

[3] Michael Burton. Android App Development For Dummies[M]. 3 版. For Dummies,2015.

[4] Google. API Guides[EB/OL]. http://developer.android.com/develop/index.html,2013.

[5] Eric Freeman. Head First 设计模式[M]. O'Reilly Taiwan 公司,译. 北京：中国电力出版社,2007.

[6] lightSky. 开源项目 Philm 的 MVP 架构分析[EB/OL]. http://www.lightskystreet.com/2015/02/10/philm_mvp/,2015.